Student Solutions Manual

Statistics for Business and Economics

Seventh Edition

Paul Newbold
William L. Carlson
Betty Thorne

Prentice Hall

Boston Columbus Indianapolis New York San Francisco Upper Saddle River

Amsterdam Cape Town Dubai London Madrid Milan Munich Paris Montreal Toronto

Delhi Mexico City Sao Paulo Sydney Hong Kong Seoul Singapore Taipei Tokyo

Editor-in-Chief: Eric Svendsen
Executive Editor: Chuck Synovec
Editorial Project Manager Mary Kate Murray
Production Project Manager: Ana Jankowski
Operations Specialist: Arnold Vila

www.pearsonhighered.com

10 9 8 7 6 5 4 3 2 1

ISBN-13: 978-0-13608538-6
ISBN-10: 013-608538-5

CONTENTS

Chapter 1:

Describing Data: Graphical

1.2
a. Categorical, nominal (The response is categorical because the responses can be grouped into classes or categories, in this case yes/no. The measurement levels are nominal because the responses are words that describe the categories.)
b. Categorical, ordinal (The response is categorical because the responses can be grouped into classes or categories. The measurement levels are ordinal because these are rankings of the data.)
c. Numerical, discrete (The response is numerical because the responses cannot be grouped into classes or categories. Since the response is an actual cost, it is discrete because the value comes from a counting process.)

1.4
a. Categorical – Qualitative – ordinal
b. Numerical – Quantitative – discrete
c. Categorical – Qualitative – nominal
d. Categorical – Qualitative – nominal

1.6
a. Categorical – Qualitative – nominal
b. Numerical – Quantitative - discrete
c. Categorical – Qualitative – nominal: yes/no response
d. Categorical – Qualitative – ordinal

1.8
a. Various answers - Categorical variable with ordinal responses: Health consciousness
b. Various answers – Categorical variable with nominal responses: Gender

1.10 Pareto diagram – possible defects for a product line

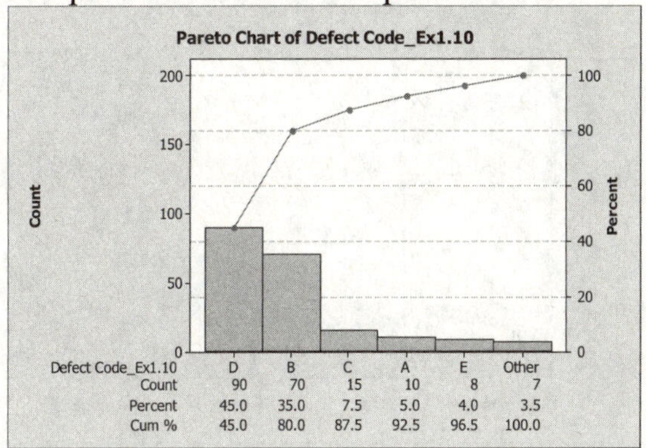

1.12

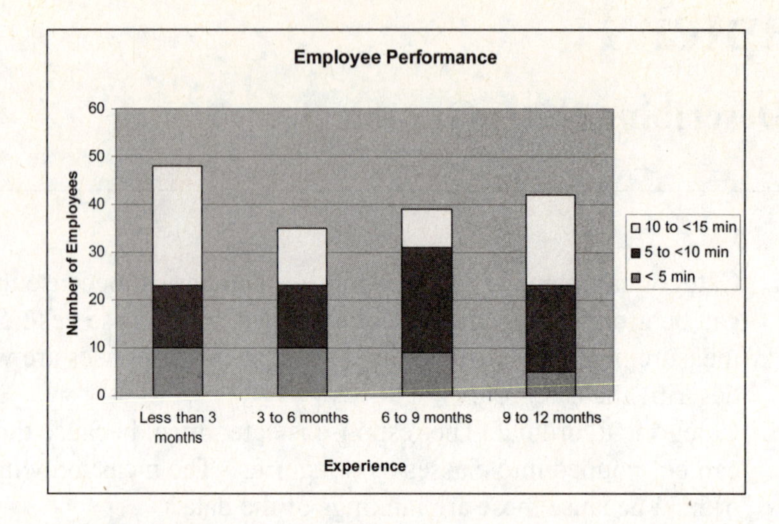

1.14 a. Bar chart of the number of endangered wildlife species in the United States

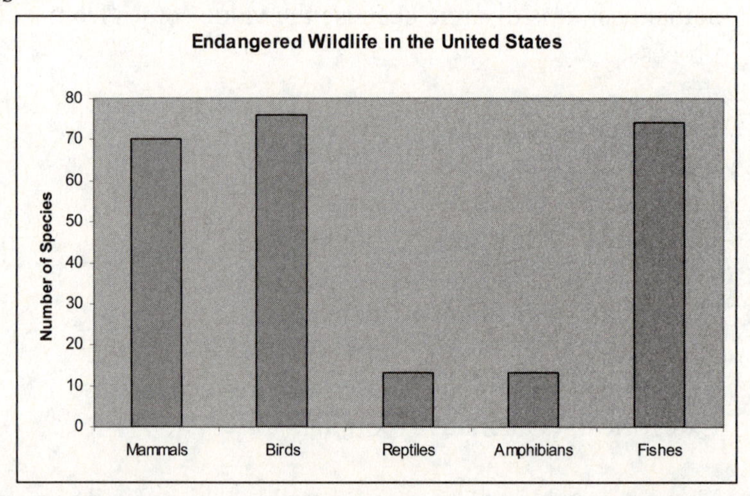

b. Bar chart of the number of endangered wildlife species outside the United States

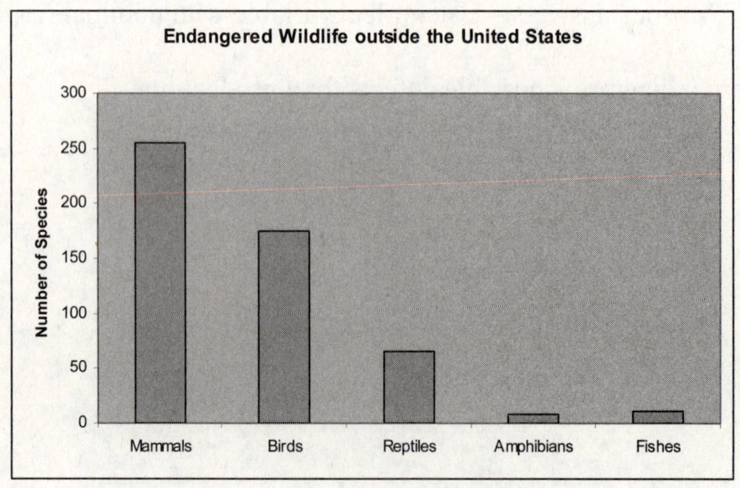

c. Bar chart to compare the number of endangered species in the United States to the number of endangered species outside the United States

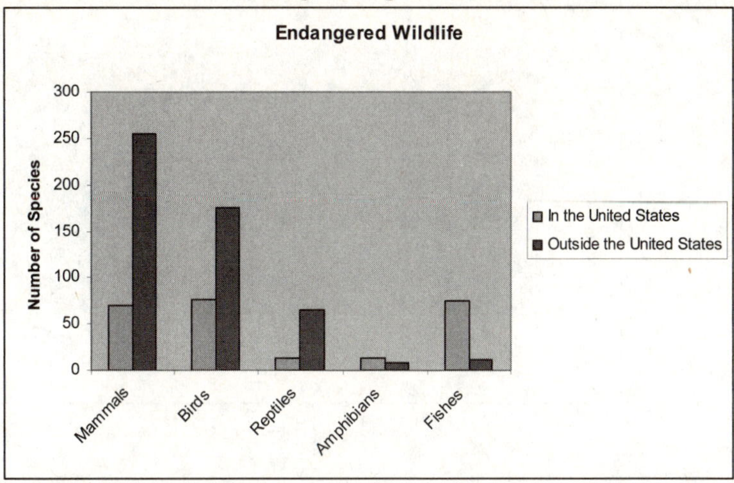

1.16 Describe the data graphically

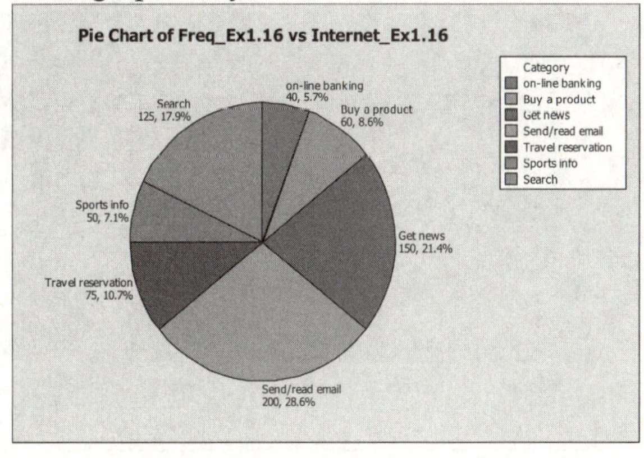

1.18 a. Component bar chart

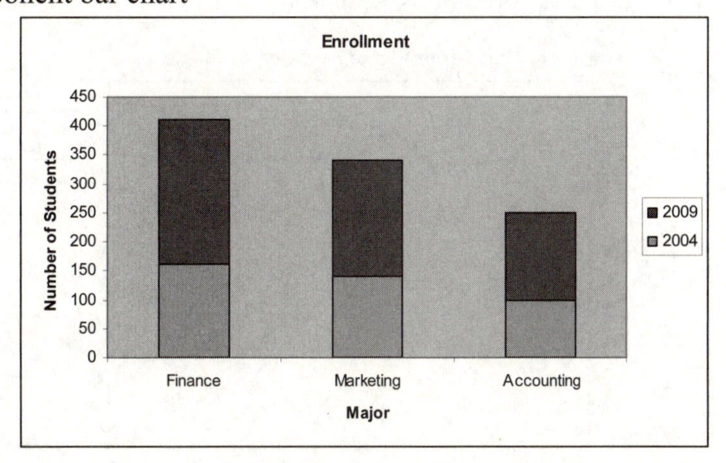

b. Cluster bar chart

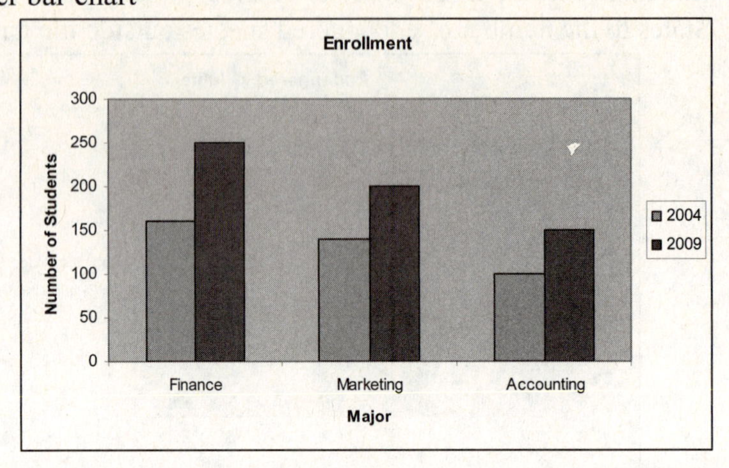

1.20

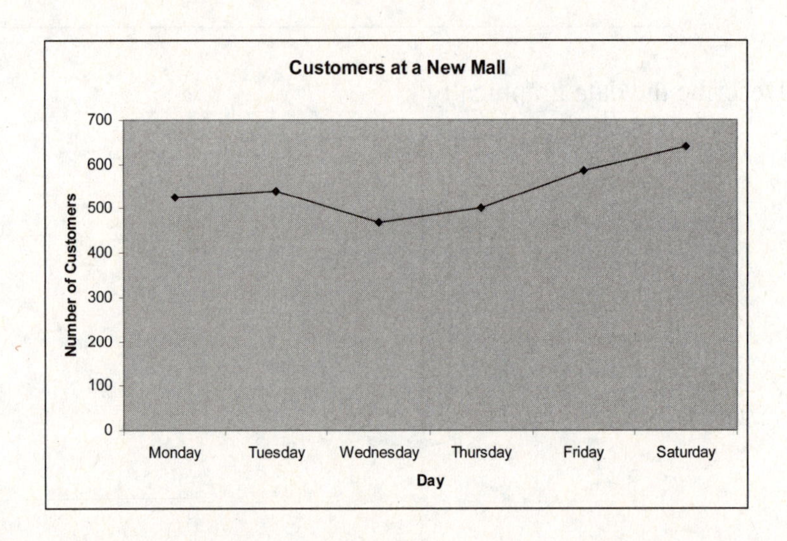

1.22 a. Time series plot of degrees awarded

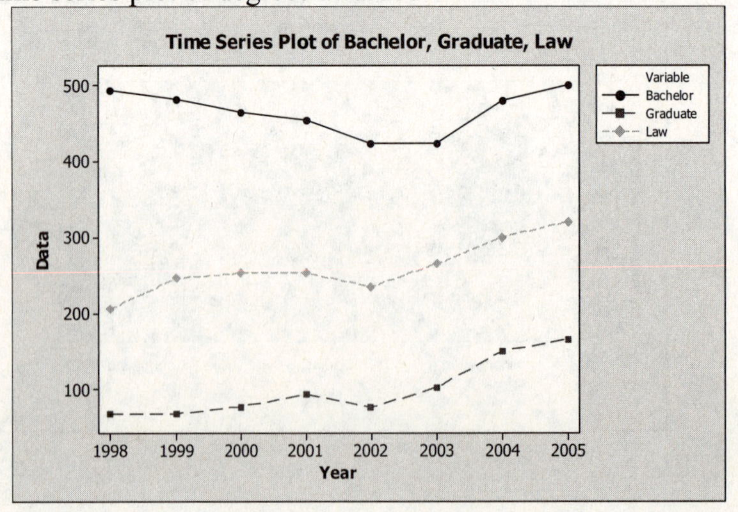

b. The number of law and graduate degrees awarded is increasing. The number of bachelor degrees awarded declined from 1998 to 2004 with a slight increase in 2005. Enrollment restrictions may be in order if class sizes are becoming too large or if crowding conditions occur.

1.24 a. The Euro (EUR) compared to 1 U.S. Dollar (USD)

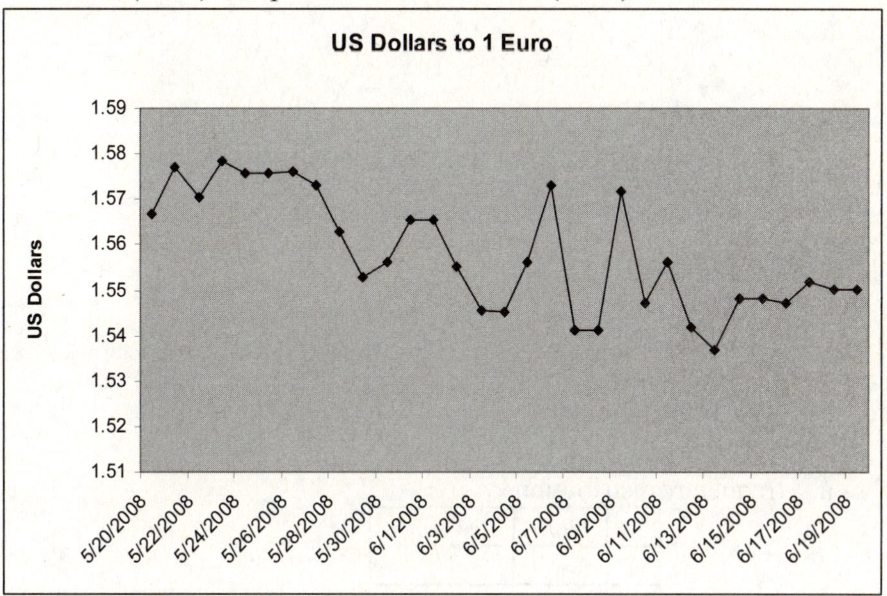

b. Answers may vary.

1.26 Tme series plot of a stock market index (Dow Jones Industrial Average) over 14 years from the internet

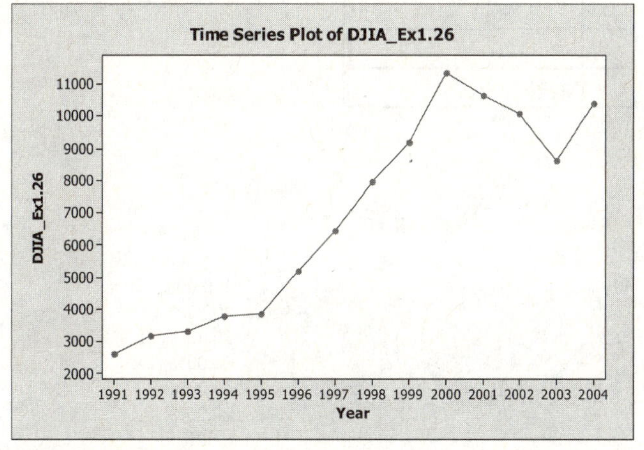

1.28 Time series plot of Housing Starts data

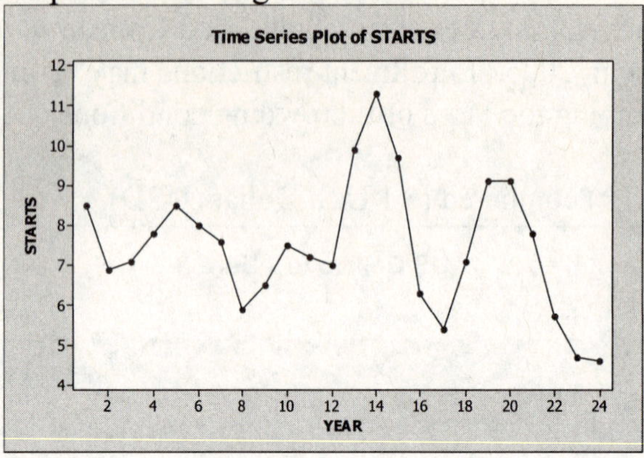

1.30
a. 5 – 7 classes
b. 7 – 8 classes
c. 8 – 10 classes
d. 8 – 10 classes
e. 10 – 11 classes

1.32
a. frequency distribution

Bin	*Frequency*
10	0
20	5
30	3
40	8
50	3
60	5
70	4
More	0

b. histogram
c. ogive

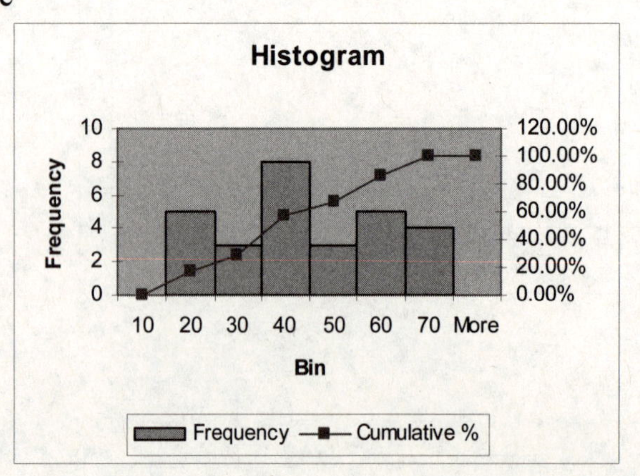

d. stem-and-leaf display

Stem-and-Leaf Display: Data_Ex1.32

Stem-and-leaf of Data_Ex1.32 N = 28
Leaf Unit = 1.0

```
2    1  23
5    1  557
7    2  14
8    2  8
9    3  2
(6)  3  567799
13   4  0144
9    4
9    5  14
7    5  699
4    6  24
2    6  55
```

1.34

Classes	Frequency	a. Relative Frequency	b. Cumulative Frequency	c. Relative Cumulative Frequency
0<10	8	16.33%	8	16.33%
10<20	10	20.41%	18	36.74%
20<30	13	26.53%	31	63.27%
30<40	12	24.49%	43	87.76%
40<50	6	12.24%	49	100.00%
Total	49	100.00%		

1.36 For the file Water - construct a frequency distribution, cumulative frequency distribution, histogram, ogive and stem-and-leaf display. Various answers – one possibility is to use 8 classes with a width of .1.

Bin	*Frequency*	*Cum Freq*	
3.5	0	0	
3.6	1	1	
3.7	8	9	
3.8	30	39	
3.9	22	61	
4	12	73	
4.1	1	74	
4.2	1	75	
More	0		

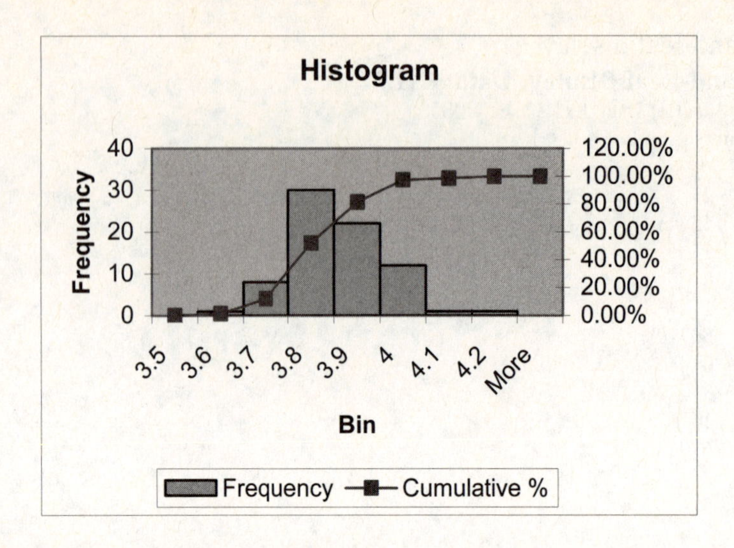

Stem-and-Leaf Display: Volume

Stem-and-leaf of Volumes N = 75
Leaf Unit = 0.010

1	35	7
3	36	34
9	36	577799
21	37	111122344444
(17)	37	55566777777889999
37	38	0111112222244
24	38	556677899
15	39	01334444
7	39	56689
2	40	
2	40	6

HI 411

11.38 a. Histogram of the Returns data

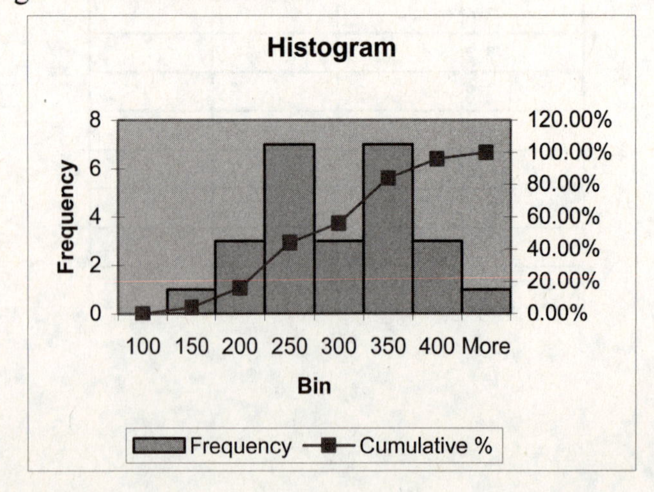

b. stem-and-leaf display

Stem-and-Leaf Display: Returns

Stem-and-leaf of Returns N = 25
Leaf Unit = 10

```
1   1 3
1   1
1   1
4   1 899
7   2 001
7   2
12  2 44445
12  2
(2) 2 89
11  3 00001
6   3 22
4   3
4   3 6
3   3 89

HI 50
```

c. ogive of the Returns data

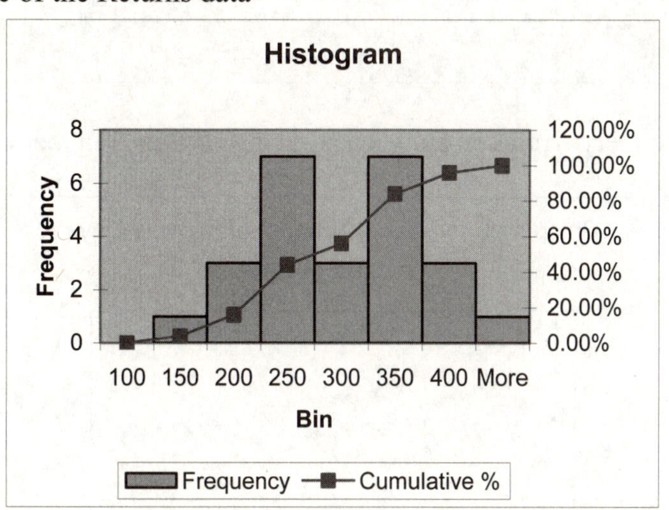

1.40 Scatterplot

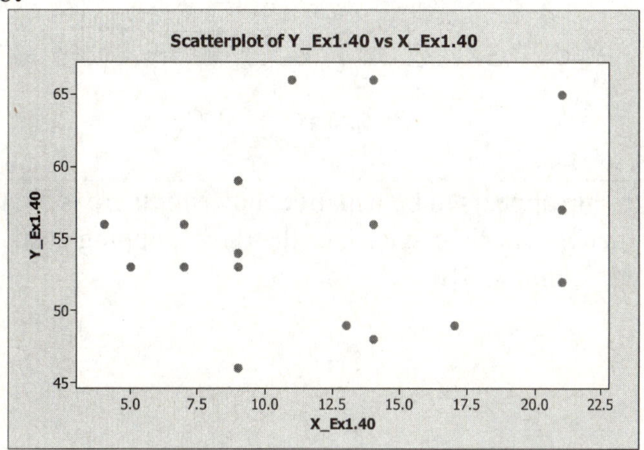

1.42 a. cross table for subcontractors and parts supplied

Subcontractor	Parts Supplied	Defective Parts
A	58	4
B	70	10
C	72	6
	200	20

b. bar chart

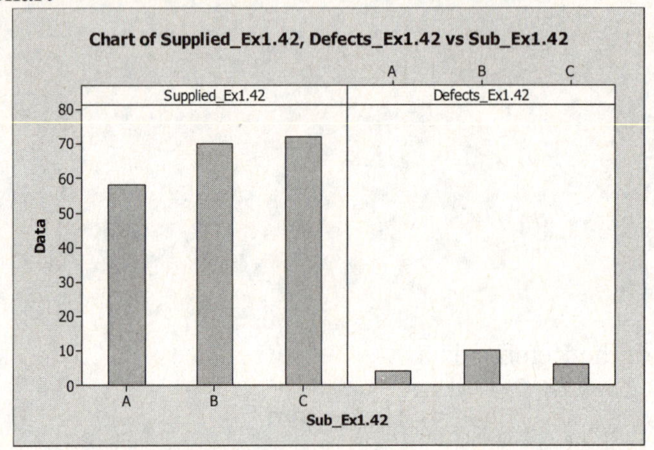

1.44 Acme Delivery – relation between shipping cost and number of delivery days

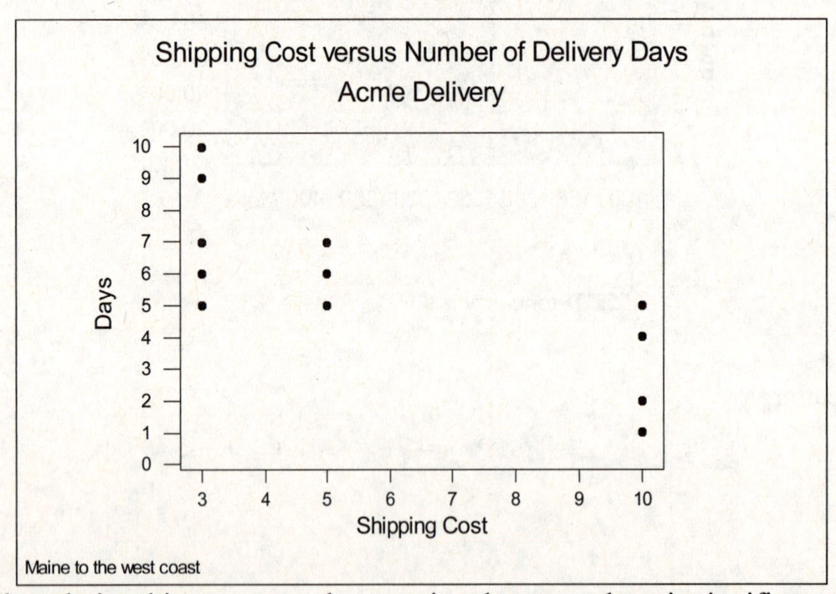

The relationship appears to be negative; however there is significant variability in delivery time at each of the three shipping costs – regular, \$3; fast, \$5; and lightning, \$10.

1.46 Scatterplot of Citydat – taxbase versus comper

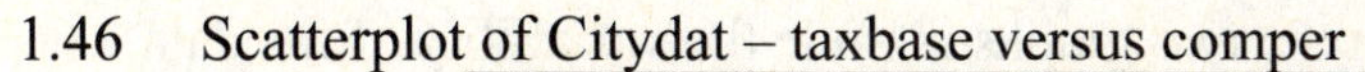

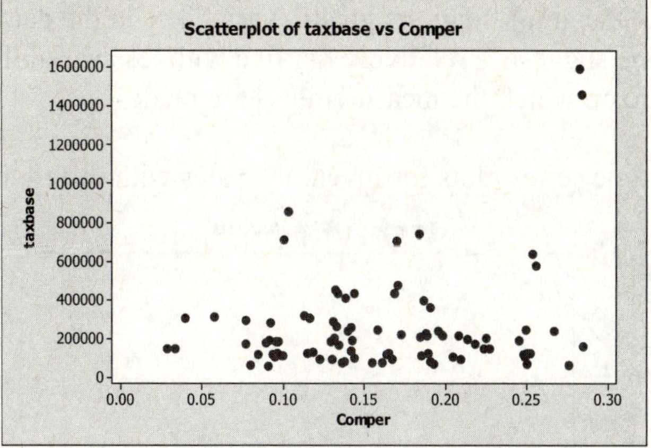

No relationship between the two variables and hence no evidence that emphasis on attracting a larger percentage of commercial property increases the tax base. The two outlier points on the right side of the plot might be used to argue that a very high amount of commercial property will provide a larger tax base. That argument, however, is contrary to the overall pattern of the data.

1.48 a. Time series plot with vertical scale from 5000 to 5700.

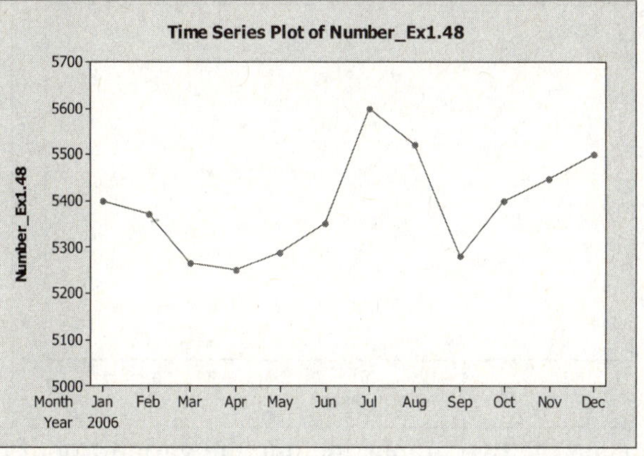

b. Time series plot with vertical scale from 4000 to 7000.

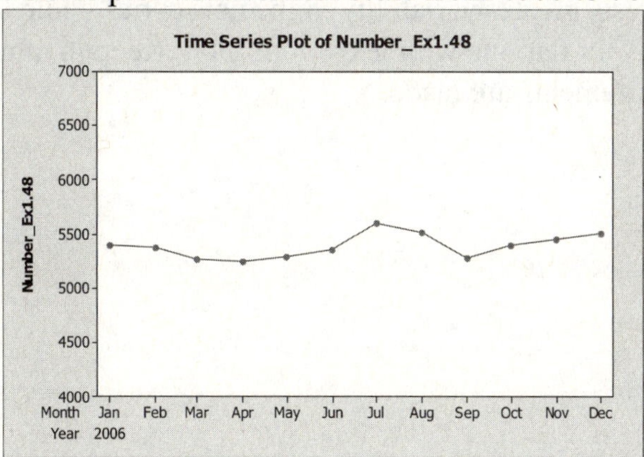

c. Differences between the two graphs include the variability of the data series. One graph suggests greater variability in the data series while the other one suggests a relatively flat line with less variability. Keep in mind the scale on which the measurements are made.

1.50 Draw two time series plots for Inventory Sales with different vertical ranges.

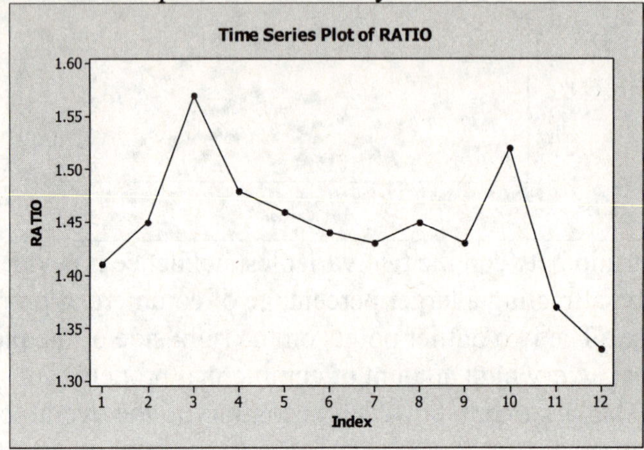

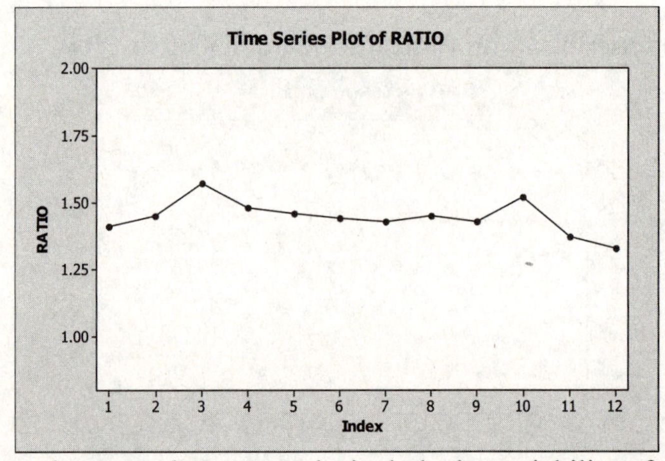

Differences between the two graphs include the variability of the data series. One graph suggests greater variability in the data series while the other one suggests a relatively flat line with less variability. Keep in mind the scale on which the measurements are made.

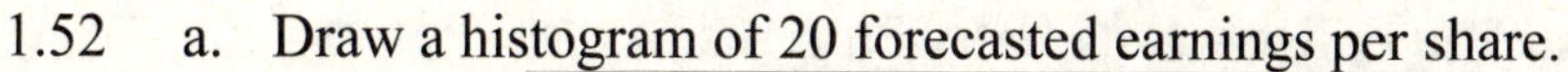

1.52 a. Draw a histogram of 20 forecasted earnings per share.

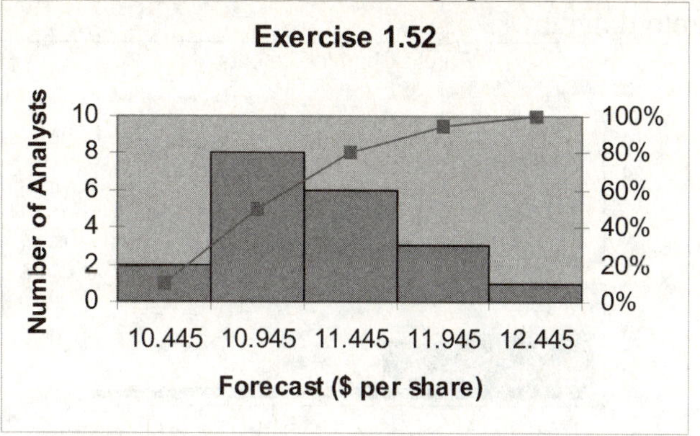

Answer to b., c. and d. are:

	(b)	(c)	(d)
Frequency	*Relative Freq.*	*Cumulative Freq.*	*Cumulative %*
2	0.1	2	10.00%
8	0.4	10	50.00%
6	0.3	16	80.00%
3	0.15	19	95.00%
1	0.05	20	100.00%

d. Cumulative relative frequencies are in the last column of the table above. These numbers indicate the percent of analysts who forecast that level of earnings per share and all previous classes, up to and including the current class. The third bin of 80% indicates that 80% of the analysts have forecasted up to and including that level of earnings per share.

1.54 Bar chart of quarterly rates of return – Stock funds:

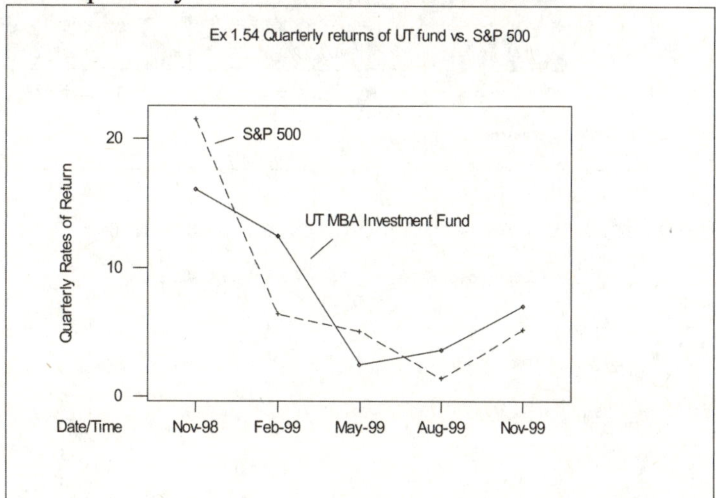

1.56 County Appraiser's Office – Data Entry Process

a. Pareto diagram

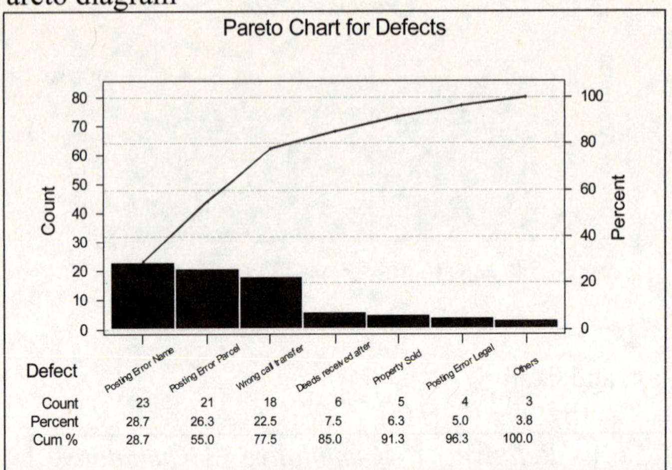

Defect	Posting Error Name	Posting Error Parcel	Wrong call transfer	Deeds received after	Property Sold	Posting Error Legal	Others
Count	23	21	18	6	5	4	3
Percent	28.7	26.3	22.5	7.5	6.3	5.0	3.8
Cum %	28.7	55.0	77.5	85.0	91.3	96.3	100.0

b. Recommendations should include a discussion of the data entry process. The data entry was being made by individuals with no knowledge of the data. Training of the data entry personnel should be a major recommendation. Increasing the size of the monitors used by the data entry staff would also reduce the number of errors.

1.58

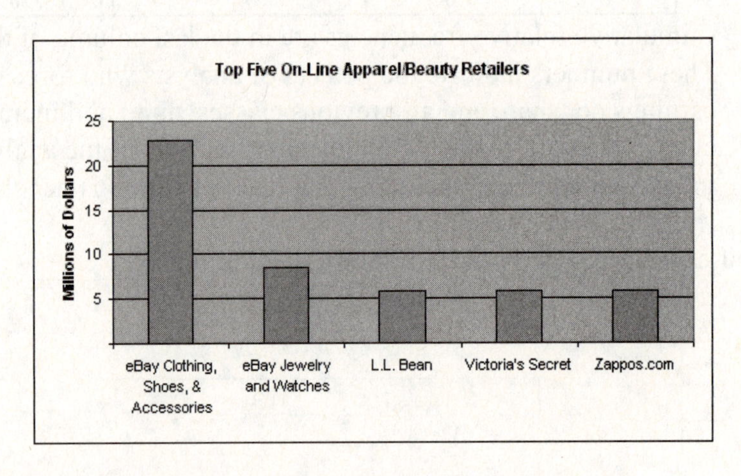

1.60 Plot the data for advertising expenditures and total sales

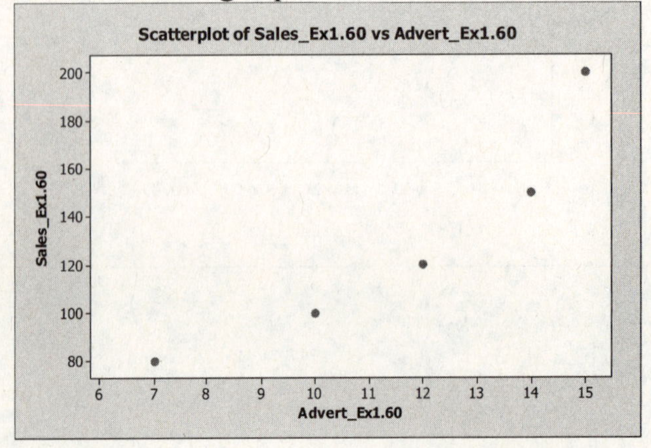

1.62 Plot the batting averages vs. hours spent per week in a weight-training program.

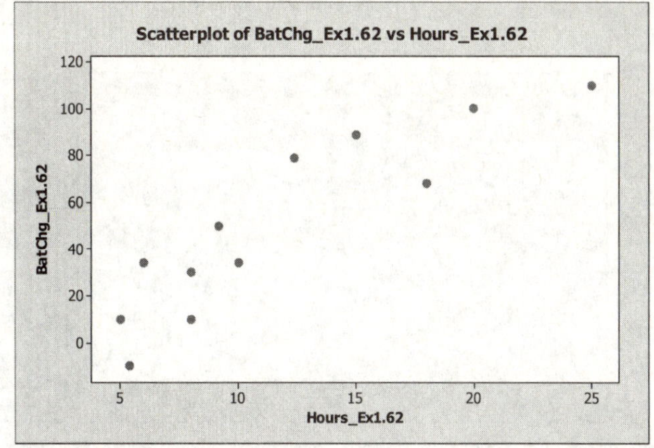

It appears that the number of hours spent per week in a special weight-training program is positively related to the change in their batting averages from the previous season.

1.64 a. Describe the new product data with a cross table

Age	Friend	Newspaper	Subtotal
<21 years	30	20	50
21-35	60	30	90
35+	18	42	60
Subtotal	108	92	200

b. Describe the data graphically

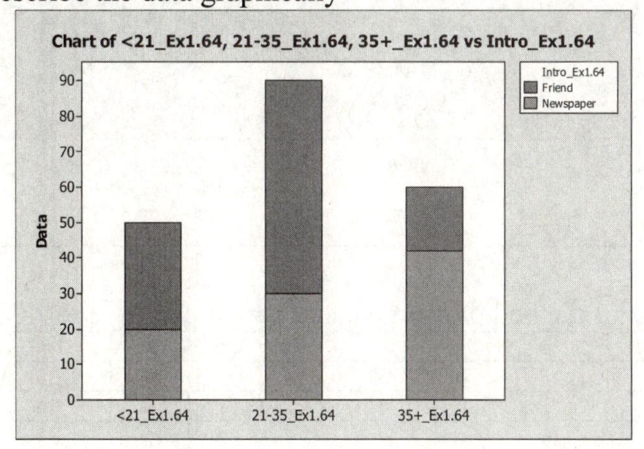

1.66 a. Time-series plot of first-year enrollments

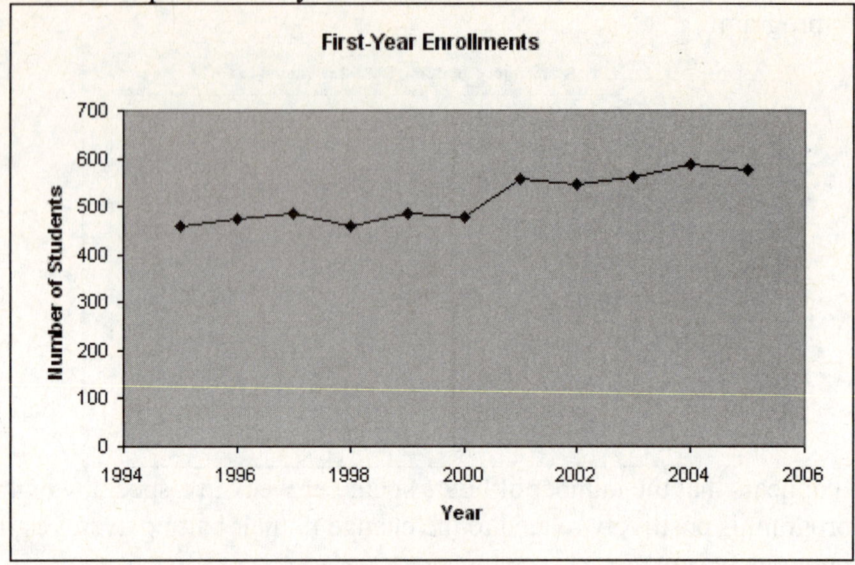

b. Time-series plot of transfer enrollments

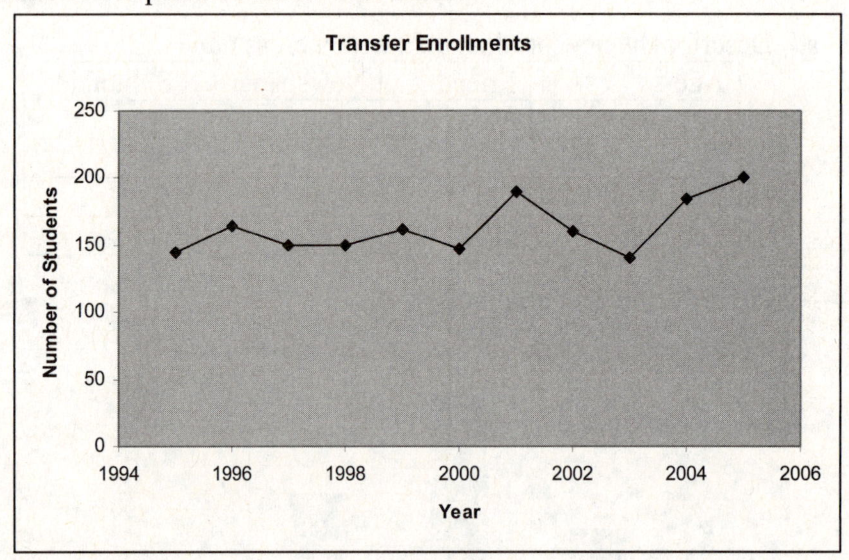

1.68 a. Cross table of method of payment and day of purchase for Florin data file.

Payment	M	T	W	Th	F	S	Tot
Am Ex	7	0	3	4	3	6	23
MC	1	4	4	2	4	9	16
Visa	6	6	4	5	8	10	24
Cash	3	1	0	0	3	9	16
Other	2	0	4	4	7	6	23
Subtotal	19	11	15	15	25	40	125

b. Pie chart of rose color preference

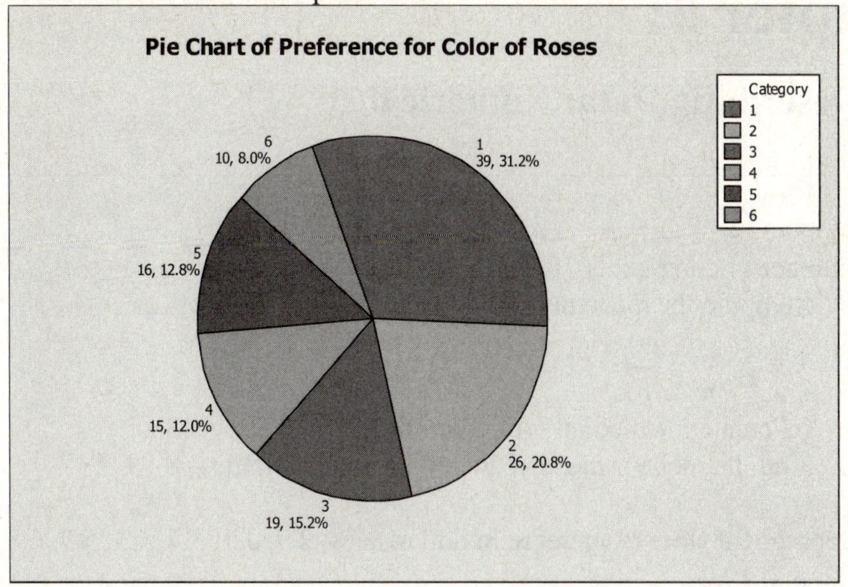

1.70

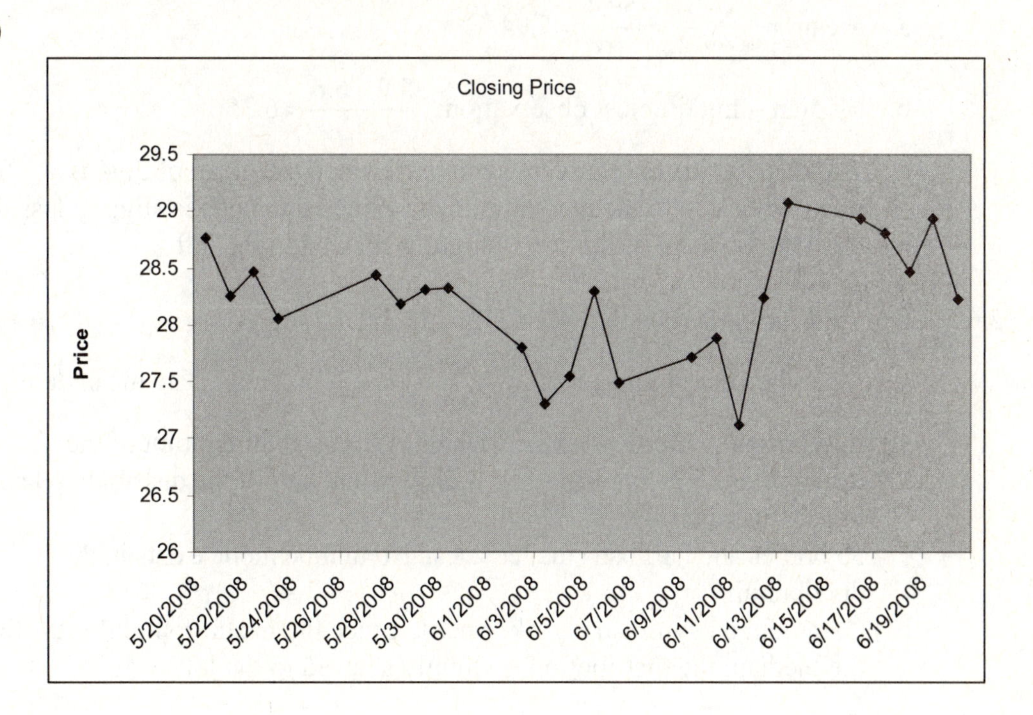

Chapter 2:

Describing Data: Numerical

2.2 Number of complaints: 8, 8, 13, 15, 16

a. Compute the mean number of weekly complaints

$$\bar{x} = \sum \frac{x_i}{n} = \frac{60}{5} = 12$$

b. Calculate the median = middlemost observation = 13

c. Find the mode = most frequently occurring value = 8

2.4 Department store % increase in dollar sales: 2.9, 3.1, 3.7, 4.3, 5.9, 6.8, 7.0, 7.3, 8.2, 10.2

a. Mean $\bar{x} = \sum \frac{x_i}{n} = \frac{59.4}{10} = 5.94$

b. Median = middlemost observation: $\frac{5.9 + 6.8}{2} = 6.35$

c. The distribution is relatively symmetric since the mean of 5.94 is relatively close to the median of 6.35. Since the mean is slightly less than the median, the distribution is slightly skewed to the left.

2.6 Demand for bottled water: 40, 43, 50, 55, 60, 62, 65

a. Describe central tendency. $\bar{x} = \sum \frac{x_i}{n} = \frac{375}{7} = 53.57$. The mean demand for one-gallon bottles is 53.57 which is the balancing point of the distribution. The median of 55 indicates that half of the distribution had larger sales than
55 bottles and half had smaller sales. No unique mode exists in the distribution

b. Comment on symmetry or skewness. Since the mean is slightly less than the median, the distribution is slightly skewed to the left.

2.8 Ages of 12 students

a. $\bar{x} = \sum \frac{x_i}{n} = \frac{307}{12} = 25.58$

b. Median = 22.50

c. Mode = 22

2.10 a. $\bar{x} = \sum \frac{x_i}{n} = \frac{282}{33} = 8.545$

b. Median = 9.0

c. The distribution is slightly skewed to the left since the mean is less than the median.

2.12 The variance and standard deviation are

$$s^2 = \frac{\sum (x_i - \bar{x})^2}{n-1} = \frac{36}{7} = 5.143 \qquad \text{and} \qquad s = 2.268$$

2.14 Calculate the coefficient of variation:

$$\bar{x} = 9; \quad s^2 = \frac{\sum (x_i - \bar{x})^2}{n-1} = \frac{10}{4} = 2.5; \quad s = 1.581;$$

CV= [1.581/9] x 100% = 17.57

2.16 a. IQR = 24.25; Q_1 = 49.5; Q_3 = 73.75

b. 8th decile = 80th percentile = 18.4th observation = 76 + 0.4(79-76) = 77.2

c. 92nd percentile = 21.26th observation = 83 + 0.16(87-83) = 83.64

2.18 Use Chebychev's theorem to approximate each of the following:

a. Between 190 and 310. +/- 3 standard deviations: At least 88.9% of the observations are within 3 standard deviations from the mean

b. Between 210 and 290. +/- 2 standard deviations: At least 75% of the observations are within 2 standard deviations from the mean

2.20 Compare the annual % returns of stocks vs. U.S. Treasury bills

Descriptive Statistics: Stocks_Ex2.20, TBills_Ex2.20

Variable	N	N*	Mean	SE Mean	TrMean	StDev	Variance	CoefVar	Minimum
Stocks_Ex2.20	7	0	8.16	8.43	*	22.30	497.39	273.41	-26.50
TBills_Ex2.20	7	0	5.786	0.556	*	1.471	2.165	25.43	3.800

Variable	Q1	Median	Q3	Maximum	Range	IQR
Stocks_Ex2.20	-14.70	14.30	23.80	37.20	63.70	38.50
TBills_Ex2.20	4.400	5.800	6.900	8.000	4.200	2.500

a. Compare the means of the populations

$$\mu_{stocks} = \frac{\sum x_i}{N} = \frac{57.12}{7} = 8.16 \quad \mu_{Tbills} = \frac{\sum x_i}{N} = \frac{40.502}{7} = 5.786$$

The mean annual % return on stocks is higher than the return for U.S. Treasury bills

b. Compare the standard deviations of the populations

$$\sigma_{stocks} = \sqrt{\sigma^2} = \sqrt{\frac{\sum (x_i - \mu)^2}{N}} = 20.648$$

$$\sqrt{\frac{(4.0-8.16)^2 + (14.3-8.16)^2 + (19-8.16)^2 + (-14.7-8.16)^2 + (-26.5-8.16)^2 + (37.2-8.16)^2 + (23.8-8.16)^2}{7}}$$

$$\sigma_{Tbills} = \sqrt{\sigma^2} = \sqrt{\frac{\sum (x_i - \mu)^2}{N}} = 1.362$$

$$\sqrt{\frac{(6.5-5.8)^2 + (4.4-5.8)^2 + (3.8-5.8)^2 + (6.9-5.8)^2 + (8.0-5.8)^2 + (5.8-5.8)^2 + (5.1-5.8)^2}{7}}$$

The variability of the U.S. Treasury bills is much smaller than the return on stocks.

2.22 a. range = 4.11 – 3.57 = 0.54, standard deviation = 0.1024, variance = 0.010486

b. Five number summary:

Min	Q1	Median	Q3	Max
3.57	3.74	3.79	3.87	4.11

c. IQR = Q3 – Q1 = 3.87 – 3.74 = .13. This tells that the range of the middle 50% of the distribution is 0.13

d. Coefficient of variation = s / $\bar{x}$ = 0.1024 / 3.8079 = 0.02689 or 2.689%

2.24 a. Standard deviation (s) of the assessment rates:

$$s = \sqrt{s^2} = \sqrt{\frac{\sum_{i=1}^{n} (x_i - \bar{x})^2}{n-1}} = \sqrt{\frac{583.75}{39}} = \sqrt{14.974} = 3.8696$$

b. The distribution is mounded. Therefore, the empirical rule applies. Approximately 95% of the distribution is expected to be within +/- 2 standard deviations of the mean.

2.26 a. mean without the weights $\bar{x} = \sum \frac{x_i}{n} = \frac{21}{5} = 4.2$

b. weighted mean

w_i	x_i	$w_i x_i$
8	4.6	36.8
3	3.2	9.6
6	5.4	32.4
2	2.6	5.2
5	5.2	26.0
24		110.0

$$\bar{x} = \frac{\sum w_i x_i}{\sum w_i} = \frac{110}{24} = 4.583$$

2.28 Find the weighted mean per capita personal income

State	Pop	Income	$w_i x_i$
AL	4,500,752	26,338	118,540,806,176
GA	8,684,715	29,442	255,695,379,030
IL	12,653,544	33,690	426,297,897,360
IN	6,195,643	28,783	178,329,192,469
NY	19,190,115	36,574	701,859,266,010
PA	12,365,455	31,998	395,669,829,090
TN	5,841,748	28,455	166,226,939,340
	69,431,972		2,242,619,309,475

$$\bar{x} = \frac{\sum w_i x_i}{\sum w_i} = \frac{2,242,619,309,475}{69,431,972} = 32,299.519$$

2.30 Based on a sample of n=50:

m_i	f_i	$f_i m_i$	$(m_i - \bar{x})$	$(m_i - \bar{x})^2$	$f_i(m_i - \bar{x})^2$
0	21	0	-1.4	1.96	41.16
1	13	13	-0.4	0.16	2.08
2	5	10	0.6	0.36	1.8
3	4	12	1.6	2.56	10.24
4	2	8	2.6	6.76	13.52
5	3	15	3.6	12.96	38.88
6	2	12	4.6	21.16	42.32
Sum	**50**	**70**			**150**

a. Sample mean number of claims per day = $\bar{X} = \frac{\sum f_i m_i}{n}$ =70/50 = 1.40

b. Sample variance = $s^2 = \frac{\sum f_i (m_i - \bar{x})^2}{n-1} = \frac{150}{49} = 3.0612$

Sample standard deviation = $s = \sqrt{s^2}$ = 1.7496

2.32 Estimate the sample mean and sample standard deviation

m_i	f_i	$f_i m_i$	$(m_i - \bar{x})$	$(m_i - \bar{x})^2$	$f_i(m_i - \bar{x})^2$
10.2	2	20.4	-0.825	0.681	1.361
10.7	8	85.6	-0.325	0.106	0.845
11.2	6	67.2	0.175	0.031	0.184
11.7	3	35.1	0.675	0.456	1.367
12.2	1	12.2	1.175	1.381	1.381
	20	220.5			5.138
x-bar=	11.03			variance =	0.2704

a. sample mean = $\bar{X} = \frac{\sum f_i m_i}{n}$ = 220.5/20 = 11.025

b. sample variance = $s^2 = \dfrac{\sum f_i(m_i - \overline{x})^2}{n-1} = 5.138/19 = 0.2704$

sample standard deviation = $s = \sqrt{s^2} = 0.520$

2.34 Calculate mean and standard deviation from Example 1.8

Minutes	m_i	f_i	$f_i m_i$	$(m_i - \overline{x})$	$(m_i - \overline{x})^2$	$f_i(m_i - \overline{x})^2$
220<230	225	5	1125	-36.545	1335.57	6677.851
230<240	235	8	1880	-26.545	704.6612	5637.289
240<250	245	13	3185	-16.545	273.7521	3558.777
250<260	255	22	5610	-6.5455	42.84298	942.5455
260<270	265	32	8480	3.45455	11.93388	381.8843
270<280	275	13	3575	13.4545	181.0248	2353.322
280<290	285	10	2850	23.4545	550.1157	5501.157
290<300	295	7	2065	33.4545	1119.207	7834.446
		110	28770			32887.27

a. $\overline{x} = \dfrac{\sum f_i m_i}{n} = \dfrac{28770}{110} = 261.54545$

b. $s^2 = \dfrac{\sum f_i(m_i - \overline{x})^2}{n-1} = \dfrac{32887.27}{109} = 301.718 \quad s = \sqrt{s^2} = 17.370$

2.36 a. compute the sample covariance

x_i	y_i	$(x_i - \overline{x})$	$(x_i - \overline{x})^2$	$(y_i - \overline{y})$	$(y_i - \overline{y})^2$	$(x_i - \overline{x})(y_i - \overline{y})$
12	200	-7	49	-156	24336	1092
30	600	11	121	244	59536	2684
15	270	-4	16	-86	7396	344
24	500	5	25	144	20736	720
14	210	-5	25	-146	21316	730
95	1780	0	236	0	133320	5570
$\overline{x} = 19.00$	$\overline{y} = 356.00$		$s_x^2 = 59$		$s_y^2 = 33330$	Cov(x,y) = 1392.5
			s_x =7.681146		s_y =182.5650569	

$$Cov(x, y) = \frac{\sum (x_i - \overline{x})(y_i - \overline{y})}{n-1} = \frac{5570}{4} = 1392.5$$

b. compute the sample correlation coefficient

$$r = \frac{Cov(x, y)}{s_x s_y} = \frac{1392.5}{(7.6811)(182.565)} = 0.9930$$

2.38 a. *Cov* (*x*,*y*) = 4.268

b. *r* = 0.128

c. Weak positive association between the number of drug units and the number of days to complete recovery. Recommend low or no dosage units.

2.40 a. Compute the covariance

x_i	y_i	$(x_i - \bar{x})$	$(x_i - \bar{x})^2$	$(y_i - \bar{y})$	$(y_i - \bar{y})^2$	$(x_i - \bar{x})(y_i - \bar{y})$
5	55	-2	4	12.4	153.76	-24.8
6	53	-1	1	10.4	108.16	-10.4
7	45	0	0	2.4	5.76	0
8	40	1	1	-2.6	6.76	-2.6
9	20	2	4	-22.6	510.76	-45.2
35	213	0	10	0	785.2	-83
$\bar{x}$ = 7.00	$\bar{y}$ = 42.60		s_x^2 = 2.5		s_y^2 = 196.3	$Cov(x, y)$ = -20.75
			s_x = 1.58114		s_y = 14.0107	

$$Cov(x, y) = \frac{\sum(x_i - \bar{x})(y_i - \bar{y})}{n-1} = \frac{-83}{4} = -20.75$$

b. Compute the correlation coefficient

$$r_{xy} = \frac{Cov(x, y)}{s_x s_y} = \frac{-20.75}{(1.58114)(14.0107)} = -.937$$

2.42 Scatter plot – Advertising expenditures (thousands of $s) vs. Monthly Sales (thousands of units)

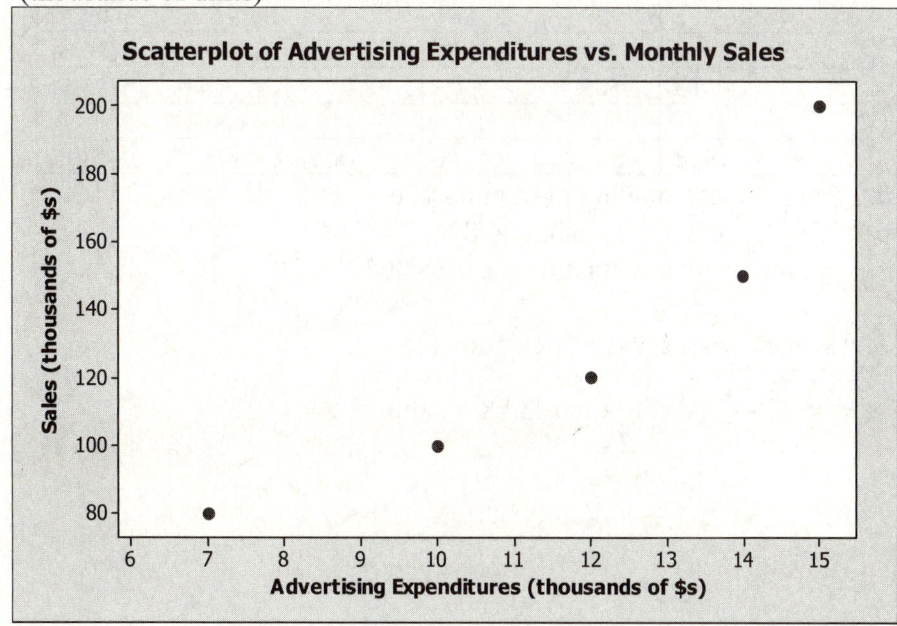

x_i	y_i	$(x_i - \bar{x})$	$(x_i - \bar{x})^2$	$(y_i - \bar{y})$	$(y_i - \bar{y})^2$	$(x_i - \bar{x})(y_i - \bar{y})$
10	100	-1.6	2.56	-30	900	48
15	200	3.4	11.56	70	4900	238
7	80	-4.6	21.16	-50	2500	230
12	120	0.4	0.16	-10	100	-4
14	150	2.4	5.76	20	400	48
58	650		41.2		8800	560
$\bar{x}$ = 11.60	$\bar{y}$ = 130.00		s_x^2 = 10.3		s_y^2 =2200	Cov(x,y) = 140
			s_x = 3.2094		s_y = 46.9042	
				b1 =	13.5922	
				b0 =	-27.67	

$$\text{Covariance} = Cov(x,y) = \frac{\sum (x_i - \bar{x})(y_i - \bar{y})}{n-1} = 560 / 4 = 140$$

$$\text{Correlation} = \frac{Cov(x,y)}{s_x s_y} = \frac{140}{(3.2094)(46.9042)} = .93002$$

2.44 Air Traffic Delays (Number of Minutes Late)

m_i	f_i	$f_i m_i$	$(m_i - \bar{x})$	$(m_i - \bar{x})^2$	$f_i(m_i - \bar{x})^2$
5	30	150	-13.133	172.46	5173.90
15	25	375	-3.133	9.81	245.32
25	13	325	6.867	47.16	613.11
35	6	210	16.867	284.51	1707.07
45	5	225	26.867	721.86	3609.30
55	4	220	36.867	1359.21	5436.84
	83	1505			16785.54
x-bar=	18.13			variance =	204.7017

a. Sample mean number of minutes late = 1505 / 83 = 18.1325

b. Sample variance = 16785.54/82 = 204.7017
sample standard deviation = s = 14.307

2.46 The variance and standard deviation are

$$s^2 = \frac{\sum (x_i - \bar{x})^2}{n-1} = \frac{1114}{9} = 123.78 \quad \text{and} \quad s = 11.13$$

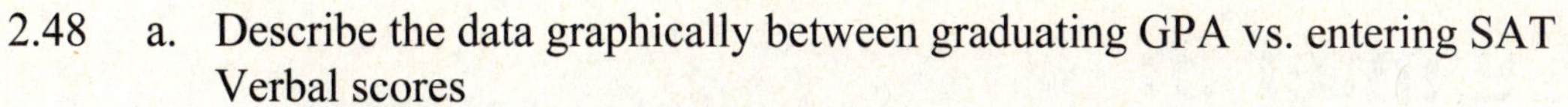

2.48 a. Describe the data graphically between graduating GPA vs. entering SAT Verbal scores

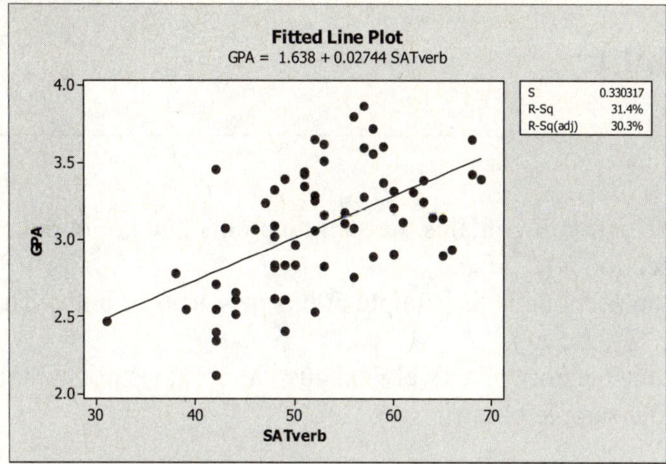

b. Describe the data numerically

Covariances: GPA, SATverb

	GPA	SATverb
GPA	0.169284	
SATverb	1.791637	65.293985

Correlations: GPA, SATverb

Pearson correlation of GPA and SATverb = 0.560

P-Value = 0.000

Regression Analysis: GPA versus SATverb

The regression equation is

GPA = 1.64 + 0.0274 SATverb

c. Estimate a graduation GPA for a student with a verbal score of 52

GPA = 1.64 + 0.0274 SATverb = 3.06

2.50 Mean of \$295 and standard deviation of \$63.

a. Find a range in which it can be guaranteed that 60% of the values lie.
Use Chebychev's theorem: at least 60% = $[1-(1/k^2)]$. Solving for k = 1.58. The interval will range from 295 +/- (1.58)(63) = 295 +/- 99.54. 195.46 up to 394.54 will contain at least 60% of the observations.

b. Find the range in which it can be guaranteed that 84% of the growth figures lie.
Use Chebychev's theorem: at least 84% = $[1-(1/k^2)]$. Solving for k = 2.5. The interval will range from 295 +/- (2.50)(63) = 295 +/- 157.5. 137.50 up to 452.50 will contain at least 84% of the observations.

2.52 Tires have a lifetime mean of 29,000 miles and a standard deviation of 3,000 miles.

a. Find a range in which it can be guaranteed that 75% of the lifetimes of tires lies
Use Chebychev's theorem: at least 75% = $[1-(1/k^2)]$. Solving for k = 2.0. The interval will range from 29,000 ± (2.0)(3000) = 29,000 ± 6000 = 23,000 to 35,000.

b. 95%, solve for K = 4.47. The interval will range from 29,000 ± (4.47)(3000) = 29,000 ± 13,416.41 = 15,583.59 to 42,416.41.

Chapter 3:

Probability

3.2 a. A intersection B contains the sample points that are in both A and B. The intersection = (E_3, E_9).
b. A union B contains the sample points in A or B or both. The union = (E_1, E_2, E_3, E_7, E_8, E_9)
c. A union B is not collectively exhaustive – it does not contain all of the possible sample points.

3.4 a. A intersection B = (E_3, E_6)
b. A union B = (E_3, E_4, E_5, E_6, E_9, E_{10})
c. A union B is not collectively exhaustive – it does not contain all of the possible sample points.

3.6 a. ($A \cap B$) is the event that the Dow-Jones average rises on both days which is O_1. ($\bar{A} \cap B$) is the event the Dow-Jones average does not rise on the first day but it rises on the second day which is O_3. The union between these two will be O_1 or O_3 either of which by definition is event B: the Dow-Jones average rises on the second day.
b. Since ($\bar{A} \cap B$) is the event the Dow-Jones average does not rise on the first day but rises on the second day which is O_3 and because A is the event that the Dow-Jones average rises on the first day, then the union will be O_2, either the Dow-Jones average does not rise on the first day but rises on the second day or the Dow-Jones average rises on the first day or both. This is the definition of $A \cup B$.

3.8 The total number of outcomes in the sample space, $N = C_2^{12} = \frac{12!}{2!(12-10)!} = 66$.

The number of ways to select 1 A from the 5 available, $C_1^5 = \frac{5!}{1!(5-1)!} = 5$.

The number of ways to select 1 B from the 7 available, $C_1^7 = \frac{7!}{1!(7-1)!} = 7$.

The number of outcomes that satisfy the condition of 1A and 1 B is 5 X 7 = 35. Therefore, the probability that a randomly selected set of 2 will include 1A and 1 B is $P_A = \frac{N_A}{N} = \frac{C_1^5 C_1^7}{C_2^{12}} = \frac{5X7}{66} = .53$

3.10 The total number of outcomes in the sample space, $N = C_4^{16} = \dfrac{16!}{4!(16-4)!} = 1,820$.

The number of ways to select 2 A's from the 10 available, $C_2^{10} = \dfrac{10!}{2!(10-2)!} = 45$.

The number of ways to select 2 B's from the 6 available, $C_2^{6} = \dfrac{6!}{2!(6-2)!} = 15$.

The number of outcomes that satisfy the condition of 2A's and 2B's is 45 X 15 = 675. Therefore, the probability that a randomly selected set of 4 will include 2A's and 2B's is $P_A = \dfrac{N_A}{N} = \dfrac{C_2^{10}C_2^{6}}{C_4^{16}} = \dfrac{45X15}{1,820} = .3709$

3.12 $\dfrac{C_2^{20,000}}{C_2^{180,000}} = \dfrac{n_A}{n} = \dfrac{20,000}{180,000} = .1111$. The probability of a random sample of 2 people from the city will contain 2 Norwegians is (.1111)(.1111) = .0123

3.14
a. $P(A) = P(10\% \text{ to } 20\% \cup \text{more than } 20\%) = .33 + .21 = .54$
b. $P(B) = P(\text{less than } -10\% \cup -10\% \text{ to } 0\%) = .04 + .14 = .18$
c. A complement is the event that the rate of return is not more than 10%
d. $P(\bar{A}) = .04 + .14 + .28 = .46$
e. The intersection between more than 10% and return will be negative is the null or empty set.
f. $P(A \cap B) = 0$
g. The union of A and B is the event that are the rates of return of; less than –10%, -10% to 0%, 0% to 10%, 10% to 20% and more than 20%.
h. $P(A \cup B) = P(A) + P(B) - P(A \cap B) = .54 + .18 - 0 = .72$
i. A and B are mutually exclusive because their intersection is the null set
j. A and B are not collectively exhaustive because their union does not equal 1

3.16 A and $\bar{A}$ of exercise 4-1. are not mutually exclusive. Since P(A) = .68 and $P(\bar{A}) = .75$, check if $P(A \cup \bar{A}) = P(A) + P(\bar{A}) = .68 + .75 = 1.41 > 1$. Therefore, if two events are not mutually exclusive, the probability of their union cannot equal the sum of their individual probabilities

3.18
a. $P(X<3) = .29 + .36 + .22 = .87$
b. $P(X>1) = .22 + .10 + .03 = .35$
c. By the third probability postulate, the sum of the probabilities of all outcomes in the sample space must sum to one.

3.20 $P(A) = .40$, $P(B) = .45$, $P(A \cup B) = .85$

By the Addition Rule, $P(A \cup B) =$
$P(A) + P(B) - P(A \cap B)$.
Therefore, $.85 = .40 + .45 - P(A \cap B)$
$P(A \cap B) = .40 + .45 - .85 = 0$

3.22 $P(A) = .60$, $P(B) = .45$, $P(A \cap B) = .30$

By the Addition Rule, $P(A \cup B) =$
$P(A) + P(B) - P(A \cap B)$.
Therefore, $P(A \cup B) = .60 + .45 - .30 = .75$

3.24 $P(A) = .80$, $P(B) = .10$, $P(A \cap B) = .08$

$$P(A \mid B) = \frac{P(A \cap B)}{P(B)} = \frac{.08}{.10} = .80$$

A and *B* are independent since the $P(A|B)$ of .80 equals the $P(A)$ of .80

3.26 $P(A) = .70$, $P(B) = .80$, $P(A \cap B) = .50$

$$P(A \mid B) = \frac{P(A \cap B)}{P(B)} = \frac{.50}{.80} = .625$$

A and *B* are not independent since the $P(A|B)$ of .625 does not equal the $P(A)$ of .70

3.28 a. $7! = 5{,}040$
b. $1 / 5{,}040 = 0.0001984$

3.30 $P_3^6 = 6!/3! = 120$. Therefore, the probability of selecting, in the correct order the three best performing stocks by chance is 1/120 = .00833

3.32 $P_3^5 = 5!/2! = 60$. Therefore, the probability of making the correct prediction by chance is 1/60 = .0167

3.34 $C_2^8 = 8!/2!6! = 28$

3.36 a. $C_2^5 = 5!/2!3! = 10$, $C_4^6 = 6!/4!2! = 15$. Since the selections areindependent, then there are (10)(15) = 150 possible combinations.
b. *P*(select a brother who is a craftsman) = $C_1^4/10 = [4!/1!3!]/10 = 4/10$. Because there are only 5 craftsmen, once a brother has been selected as a craftsman there are only four ways to fill the second craftsman spot on the work crew. *P*(select a brother who is a laborer) = $C_3^5/15 = [5!/2!3!]/10$ = 10/15. Multiply the two probabilities together to find their intersection: (4/10)(10/15) = .2667

c. The probability of the complements is 1 minus the probability of the event. Therefore, P(not selecting a brother who is a craftsman = 1 – 4/10 = 6/10. P(not selecting a brother who is a laborer) = 1-10/15 = 5/15. Multiply the two probabilities together to find their intersection: (6/10)(5/15) = .20

3.38 Let A – employment concern, B – grade concerns, $A \cap B$ – both. Then $P(A \cup B) = P(A) + P(B) - P(A \cap B) = .30 + .25 - .20 = .35$

3.40 a. No, the two events are not mutually exclusive because $P(A \cap B) \neq 0$.
b. No, the two events are not collectively exhaustive because $P(A \cup B) \neq 1$
c. No, the two events are not statistically independent because $P(A \cap B) = .15 \neq .6 = P(A)P(B)$

3.42 $P(A \cup B \cup C) = P(A) + P(B) + P(C) - P(A \cap B) - P(A \cap C) - P(B \cap C) = .02 + .01 + .04 - .0002 - .0008 - .000 = .069$

3.44 Let A – watch a TV program oriented to business and financial issues, B –read a publication, then the $P(A) = .18$, $P(B) = .12$ and $P(A \cap B) = .10$
a. Find $P(B|A) = P(A \cap B) / P(A) = .10/.18 = .5556$
b. Find $P(A|B) = P(A \cap B) / P(B) = .10/.12 = .8333$

3.46 The number of ways of randomly choosing 2 stocks in order out of 4 is: $P_2^4 = 4!/2! = 12$, the number of ways of randomly choosing 2 bonds in order from 5: $P_2^5 = 5!/3! = 20$. Then the probability of choosing either the stocks in order or the bonds in order is the union between the two events which is equal to the sum of the individual probabilities minus the probability of the intersection. = 1/12 + 1/20 – 1/240 = .1292

3.48 Let event A—portfolio management was attended, B – Chartism attended, C – random walk attended, then $P(A) = .4$, $P(B) = .5$, $P(C) = .80$.
a. Find $P(A \cup B) = .4 + .5 - 0 = .9$
b. Find $P(A \cup C)$ if A and C are independent events = .4 + .8 – .32 = .88
c. If the $P(C|B) = .75$, then $P(B \cap C) = P(C|B)P(B) = (.75)(.5) = .375$.
$P(C \cap B) = P(C) + P(B) - P(C \cap B) = .8 + .5 - .375 = .925$

3.50 Let A – work related problem occurs on Monday and B – work related problem occrs in the last hour of the day's shift, then $P(A) = .3$, $P(B) = .2$ and $P(A \cap B) = .04$, $P(A \cap \bar{B}) = P(A) - P(A \cap B) = .3 - .04 = .26$
a. $P(\bar{B}|A) = P(A \cap \bar{B})/P(A) = .26/.3 = .867$
b. Check if $P(A \cap B) = P(A)P(B)$ Since $.04 \neq .06$, the two events are not independent events

3.52 Let A – new customer, B – call to a rival service customer, then $P(A) = .15$, $P(B) = .6$ and $P(B|A) = .8$. $P(A|B) = P(A \cap B / P(B)$ where $P(A \cap B) = P(B|A)P(A)$. $[(.8)(.15)]/.6 = .2$

3.54 $P(\text{High Income} \cap \text{Never}) = .05$

3.56 $P(\text{Middle Income} \cap \text{Never}) = .05$

3.58 $P(\text{High Income}|\text{Never}) = \dfrac{P(High\ Income \cap Never)}{P(Never)} = \dfrac{.05}{.30} = .1667$

3.60 $P(\text{Regular}|\text{High Income}) = \dfrac{P(Regular \cap High)}{P(High)} = \dfrac{.10}{.25} = .40$

3.62 $\text{Odds} = \dfrac{.5}{1-.5} = 1$ to 1 odds

3.64 $P(\text{High scores}|{>}25 \text{ hours}) = .40$, $P(\text{Low scores}|{>}25) = .20$, $\dfrac{.4}{.2} = 2.00$.

Studying increases the probability of achieving high scores.

3.66 Let F – frequent, I – Infrequent, O – Often, S – Sometimes and N- Never.

a. $P(F \cap O) = .12$
b. $P(F|N) = P(F \cap N)/P(N) = .19 / .27 = .7037$
c. Check if $P(F \cap N) = P(F)P(N)$. Since $.19 \neq .2133$, the two events are not independent
d. $P(O|I) = P(I \cap O) / P(I) = .07/.21 = .3333$
e. Check if $P(I \cap O) = P(I)P(O)$. Since $.07 \neq .0399$, the two events are not independent
f. $P(F) = .79$
g. $P(N) = .27$
h. $P(F \cup N) = P(F) + P(N) - P(F \cap N) = .79 + .27 - .19 = .87$

3.68 Let A – Regularly read business section, B – Occasionally, C – Never, TS – Traded stock

a. $P(C) = .25$
b. $P(TS) = .32$
c. $P(TS|C) = P(TS \cap C)/P(C) = .04/.25 = .16$
d. $PC|TS) = P(TS \cap C / P(TS) = .04/.32 = .125$
e. $P(TS|\bar{A}) = P(TS \cap (B \cup C)/PB \cup C) = (.10 + .04)/(.41+.25) = .2121$

3.70 Let Y – problems were worked, N – Problems not worked

a. $P(\text{Y}) = .32$

b. $P(A) = .25$

c. $P(A|\text{Y}) = P(A \cap \text{Y})/P(\text{Y}) = .12/.32 = .375$

d. $P(\text{Y}|A) = P(A \cap \text{Y})/P(A) = .12/.25 = .48$

e. $P(\text{C} \cup \text{belowC}|\text{Y}) = P(\text{C} \cup \text{below C} \cap \text{Y})/P(\text{Y}) = (.12 + .02)/.32 = .4375$

f. No, since $P(A \cap \text{Y})$ which is $.12 \neq P(A)P(\text{Y})$ which is .08

3.72 Let R – Readers, V – voted in the last election

a. $P(\text{V}) = .76$

b. $P(R) = .77$

c. $P(\overline{V}\,|\,\overline{R}) = P(\overline{V} \cap \overline{R})/P(\overline{R}) = .1/.23 = .4348$

3.74 $P(\text{G}|\text{H}) = .1$, $P(\text{G}|\text{L}) = .8$, $P(\text{G}|\text{S}) = .5$, $P(\text{H}) = .25$, $P(\text{L}) = .15$, $P(\text{S}) = .6$

a. $P(\text{G} \cap \text{H}) = P(\text{G}|\text{H})P(\text{H}) = (.1)(.25) = .025$

b. $P(\text{G}) = P(\text{G} \cap \text{H}) + P(\text{G} \cap \text{L}) + P(\text{G} \cap \text{S}) = .025 + (.8)(.15) + (.5)(.6) = .445$

c. $P(\text{L}|\text{G}) = P(\text{G} \cap \text{L}) / P(\text{G}) = (.8)(.15)/.445 = .2697$

3.76 $P(10\%|\text{T}) = .7$, $P(10\%|\text{M}) = .5$, $P(10\%|B) = .2$

a. $P(10\%) = P(10\% \cap \text{T}) + P(10\% \cap \text{M}) + P(10\% \cap B) = (.7)(.25) + (.5)(.5) + (.2)(.25) = .475$

b. $P(\text{T}|10\%) = P(10\% \cap \text{T})/P(10\%) = (.7)(.25)/.475 = .3684$

c. $P(\overline{T}\,|\,\overline{10\%}) = P(\overline{10\%} \cap \overline{T})/P(\overline{10\%}) = P(\overline{10\% \cup T}) \,/\, P(\overline{10\%}) =$
$[1 - P(10\% \cup \text{T})]/P(\overline{10\%}) = [1 - (.475 + .25 - (.7)(.25))] / .525 = .8571$

3.78 Let M – faulty machine, I – impurity

$P(\text{M}) = .4$, $P(\text{I}|\text{M}) = .1$, $P(\text{I}) = P(\text{I} \cap \text{M}) + P(\text{I} \cap \overline{M}) = (.4)(.1) + 0 = .04$

$P(\text{M}|\overline{I}) = P(\overline{I} \cap \text{M}) / P(\overline{I}) = [P(\text{M}) - P(\text{I}\overline{I}\,\text{M})] / P(\overline{I}) = (.4 - .04)/.96 = .375$

3.80 $P(A_1) = .4$, $P(B_1|A_1)=.6$, $P(B_1|A_2) = .7$

Complements: $P(A_2) = .6$, $P(B_2|A_1)=.4$, $P(B_2|A_2) = .3$

$$P(A_1 \mid B_1) = \frac{P(B_1 \mid A_1)P(A_1)}{P(B_1 \mid A_1)P(A_1) + P(B_1 \mid A_2)P(A_2)} = \frac{.6(.4)}{.6(.4) + .7(.6)} = .3636$$

3.82 $P(A_1) = .5$, $P(B_1|A_1)=.4$, $P(B_1|A_2) = .7$

Complements: $P(A_2) = .5$, $P(B_2|A_1)=.6$, $P(B_2|A_2) = .3$

$$P(A_1 \mid B_2) = \frac{P(B_2 \mid A_1)P(A_1)}{P(B_2 \mid A_1)P(A_1) + P(B_2 \mid A_2)P(A_2)} = \frac{.6(.5)}{.6(.5) + .3(.5)} = .6667$$

3.84 $P(A_1) = .6$, $P(B_1|A_1)=.6$, $P(B_1|A_2) = .4$
Complements: $P(A_2) = .4$, $P(B_2|A_1)=.4$, $P(B_2|A_2) = .6$

$$P(A_1 \mid B_1) = \frac{P(B_1 \mid A_1)P(A_1)}{P(B_1 \mid A_1)P(A_1) + P(B_1 \mid A_2)P(A_2)} = \frac{.6(.6)}{.6(.6)+.4(.4)} = .6923$$

3.86 E_1 = Stock performs much better than the market average
E_2 = Stock performs same as the market average
E_3 = Stock performs worse than the market average
A = Stock is rated a 'Buy'
Given that $P(E_1) = .25$, $P(E_2) = .5$, $P(E_3) = .25$, $P(A| E_1) = .4$, $P(A| E_2) = .2$, $P(A| E_3) = .1$
Then, $P(E_1 \cap A) = P(A| E_1)P(E_1) = (.4)(.25) = .10$
$P(E_2 \cap A) = P(A| E_2)P(E_2) = (.2)(.5) = .10$
$P(E_3 \cap A) = P(A| E_3)P(E_3) = (.1)(.25) = .025$

$$P(E_1 \mid A) = \frac{P(A \mid E_1)P(E_1)}{P(A \mid E_1)P(E_1) + P(A \mid E_2)P(E_2) + P(A \mid E_3)P(E_3)} =$$

$$= \frac{(.40)(.25)}{(.4)(.25)+(.2)(.5)+(.1)(.25)} = .444$$

3.88 Mutually exclusive events are events such that if one event occurs, the other event cannot occur. For example, a U.S. Senator voting in favor of a tax cut cannot also vote against it. Independent events are events such that the occurrence of one event has no effect on the probability of the other event. For example, whether or not you ate breakfast this morning is unlikely to have any effect on the probability of a U.S. Senator voting in favor of a tax cut.

3.90 Conditional probability is understanding the probability of one event given that another event has occurred. This is utilizing prior information that a specific event has occurred and then analyzing how that impacts the probability of another event. The importance is incorporating information about a known event. This has the impact of reducing the sample space in an experiment.

3.92
a. True: By definition, probabilities cannot be negative, hence the union between two events is $P(A) + P(B) - P(A \cap B)$ must be ≥ 0. Adding $P(A \cap B)$ to both sides, it follows that $P(A) + P(B)$ must be $\geq P(A \cap B)$
b. True: this follows from the probability of the union between two events. $P(A) + P(B) – P(A \cap B)$ cannot be larger than $P(A) + P(B)$. Only if the term $P(A \cap B)$ is negative can the statement be false. Since probabilities of events must be between 0 and 1 inclusive, the union between two events cannot be larger than the sum of the individual probabilities.
c. True: events cannot intersect with another event by more than their individual size

d. True: By definition of a complement, if an event occurs, then the event of 'not the event' (its complement) cannot occur.
e. False: Probabilities of any two events can sum to more than 1. Among other conditions, this statement will be true if the two events are mutually exclusive
f. False: there may be more events contained in the sample space. Hence, if two events are mutually exclusive, they may or may not be collectively exhaustive
g. False: the two events could contain common elements and hence their intersection would not be zero

3.94 a. False: Given that event B occurs has changed the sample space, hence, the revised probability may be less
b. False: If the probability of its complement is zero, then the event and its complement will be dependent events
c. True: This follows because the probabilities of events must be non-zero. By the Multiplicative Law of Probability: $P(A|B)P(B) = P(A \cap B)$. Since $0 \leq P(B) \leq 1$, then the $P(A|B) \geq P(A \cap B)$.
d. False: this statement is true only when the two events are independent
e. False: the posterior probability of any event could be smaller, larger, or equal to the prior probability

3.96 Let W—weather condition caused the accident, BI—bodily injuries were incurred. P(W) = .3, P(BI) = .2, P(W|BI) = .4
a. $P(W \cap BI) = P(W|BI)P(BI) = (.4)(.2) = .08$
b. No, check if $P(W \cap BI) = .08 \neq .06 = P(W)P(B)$
c. $P(BI|W) = P(W \cap BI)/P(W) = .08/.3 = .267$
d. $P(\overline{W} \cap \overline{BI}) = P(\overline{W \cup B}) = 1\text{-}P(W \cup BI) = 1 - P(W) - P(BI) + P(W \cap BI) =$ 1-.3-.2+.08 = .58

3.98 Let D – students drink at least once a week, B – student has a B average or better. P(D) = .35, $P(B)$ = .4, $P(B|$D) = .3
a. $P(B \cap D) = P(B|D)P(D) = (.3)(.35) = .105$
b. $P(D|B) = P(B \cap D)/P(B) = .105/.4 = .2625$
c. $P(B \cup D) = P(B) + P(D) - P(B \cap D) = .4 + .35 - .105 = .645$
d. $P(\overline{D}|\overline{B}) = P(\overline{B} \cap \overline{D})/P(\overline{B}) = [1\text{-}P(B \cup D)]/P(\overline{B}) = .355/.6 = .5917$
e. No, because $P(B \cap D)$ which is .105 ≠ .14 which is $P(B)P(D)$
f. No, their intersection is not zero, hence the two events cannot be mutually exclusive
g. No, the sum of their individual probabilities is .645 which is less than 1

3.100 Let 160 – farm size exceeds 160 acres, 50 – farm owner is over 50 years old.
$P(160) = .2$, $P(50) = .6$, $P(50|160) = .55$

a. $P(160 \cap 50) = P(50|160)P(160) = (.55)(.2) = .11$
b. $P(160 \cup 50) = P(160) + P(50) - P(160 \cap 50) = .2 + .6 - .11 = .69$
c. $P(160|50) = P(160 \cap 50)/P(50) = .11/.6 = .1833$
d. No, check if $P(160 \cap 50)$ which is $.11 \neq .12$ which is $P(160)P(50)$

3.102 Let NS – night shift worker, F- women, M – men, FP – favored plan.
$P(NS) = .5$, $P(F) = .3$, $P(M) = .7$, $P(F|NS) = .2$, $P(M|NS) = .8$, $P(FP|NS) = .65$, $P(FP|F) = .4$. Therefore, $P(NS \cap F) = (.5)(.2) = .1$, $P(FP) = P(FP|M)P(M) + P(FP|F)P(F) = (.5)(.7) + (.4)(.3) = .47$, $P(NS \cap FP) = P(FP|NS)P(NS) = (.65)(.5) = .325$

a. $P(FP \cap F) = P(FP|F)P(F) = (.4)(.3) = .12$
b. $P(NS \cup F) = P(NS) + P(F) - P(NS \cap F) = .5 + .3 - .1 = .7$
c. No, check if $P(NS \cap F)$ which is $.1 \neq .15$ which is $P(NS)P(F)$
d. $P(NS|F) = P(NS \cap F)/P(F) = .1/.3 = .3333$
e. $P(\overline{NS} \cap \overline{FP}) = 1 - P(NS \cup FP) = 1 - P(NS) - P(FP) + P(NS \cap FP) = 1 - .5 - .47 + .325 = .355$

3.104 a. $C_2^{12} = 12!/2!10! = 66$
b. $P(\text{faulty}) = C_1^{11}/66 = 11/66 = .1667$

3.106 a. $P(R_J) = P(R_J|G_J)P(G_J) + P(R_J|PR_J)P(PR_J) + P(R_J|DM_J)P(DM_J) + P(R_J|SS_J)P(SS_J) = (.81)(.08) + (.79)(.41) + (.6)(.06) + (.21)(.45) = .5192$
b. $P(R_F) = P(R_F|G_F)P(G_F) + P(R_F|PR_F)P(PR_F) + P(R_F|DM_F)P(DM_F) + P(R_F|SS_F)P(SS_F) = (.8)(.1) + (.76)(.57) + (.51)(.24) + (.14)(.09) = .6482$
c. The probability of renewals in February have increased; however, the conditional probability of renewal fell in each of the categories

3.108 $P(S|P) = P(S \cap P)/P(P) = P(P|S)P(S)/[P(P|S)P(S) + P(P|\overline{S})P(\overline{S})] = (.7)(.4)/[(.7)(.4) + (.5)(.6)] = .4828$

3.110 a. $P(W|FW) = P(W \cap FW) / P(FW) = .149/.29 = .5138$
b. $P(\overline{I}|FI) = P(\overline{I} \cap FI)/P(FI) = .181/.391 = .4629$

3.112 a. $P(G) = P(G|S)P(S) + P(G|\overline{S})P(\overline{S}) = (.7)(.6) + (.2)(.4) = .5$
b. $P(S|G) = P(G \cap S)/P(G) = .42/.5 = .84$
c. No, since $P(G \cap S)$ which is $.42 \neq .3$ which is $P(G)P(S)$
d. $P(S \geq 1) = 1 - P(S = 0) = 1 - .4^5 = .9898$

3.114 a. $P(\text{P}) = P(\text{P}|\text{HS})P(\text{HS}) + P(\text{P}|\text{C})P© + P(\text{P}|\text{O})P(\text{O}) = (.2)(.3)+(.6)(.5) + (.8)(.2) = .52$

b. $P(\overline{HS}|\text{P}) = P((\text{C}\cup\text{O})\cap\text{P})/P(\text{P}) = [P(\text{C}\cap\text{P}) + P(\text{P}\cap\text{O})]/P(\text{P})$
$= (.3 + .16)/.52 = .46/.52 = .8846$

3.116 Let R_1 – Sally is guilty, R_2 – Sally is not guilty. Let G_1 – wearing gloves, G_2 – not wearing gloves. $P(R_1) = 0.50$, $P(R_2) = 0.50$ and $P(G_1|R_1) = .60$, $P(G_1|R_2) = .80$, $P(G_2|R_1) = .40$, $P(G_2|R_2) = .20$

a. $$P(R_1|G_1) = \frac{P(G_1 | R_1)P(R_1)}{P(G_1 | R_1)P(R_1) + P(G_1 | R_2)P(R_2)} = \frac{(.60)(.50)}{(.60)(.50) + (.80)(.50)} = 0.43$$

b. Not likely since wearing gloves reduces her probability

3.118 Let G_1 – Sales will grow, G_2 – Sales will not grow. Let N_1 – new operating system, N_2 – no new operating system. $P(G_1) = 0.70$, $P(G_2) = 0.30$ and $P(N_1|G_1) = .30$, $P(N_1|G_2) = .10$, $P(N_2|G_1) = .70$, $P(N_2|G_2) = .90$

$$P(G_1 | N_1) = \frac{P(N_1 | G_1)P(G_1)}{P(N_1 | G_1)P(G_1) + P(N_1 | G_2)P(G_2)} = \frac{(.30)(.70)}{(.30)(.70) + (.10)(.30)} = 0.875$$

Chapter 4:

Discrete Random Variables and Probability Distributions

4.2 The number of defective parts produced in daily production is a discrete random variable that can take on no more than a countable number of values.

4.4 Discrete random variable – number of plays is countable

4.6 Total sales, advertising expenditures, sales of competitors

4.8 Discrete – the number of purchases is a countable number of values

4.10 Probability distribution of number of heads in one toss

X-number of heads	P(x)
0	.5
1	.5

4.12 Various answers

X –# of times missing class	P(x)	F(x)
0	.65	.65
1	.15	.80
2	.10	.90
3	.09	.99
4	.01	1.00

4.14 a. Cumulative probability function:

X	0	1	2	3	4	5	6	7	8	9
P(x)	.10	.08	.07	.15	.12	.08	.10	.12	.08	.10
F(x)	.10	.18	.25	.40	.52	.60	.70	.82	.90	1.00

b. $P(x \geq 5) = .08 + .10 + .12 + .08 + .10 = .48$

c. $P(3 \leq x \leq 7) = .15 + .12 + .08 + .10 + .12 = .57$

4.16 a. Probability distribution function

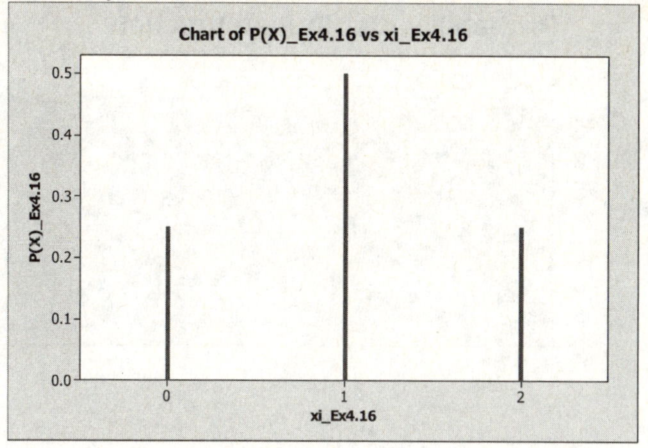

b. Cumulative probability function

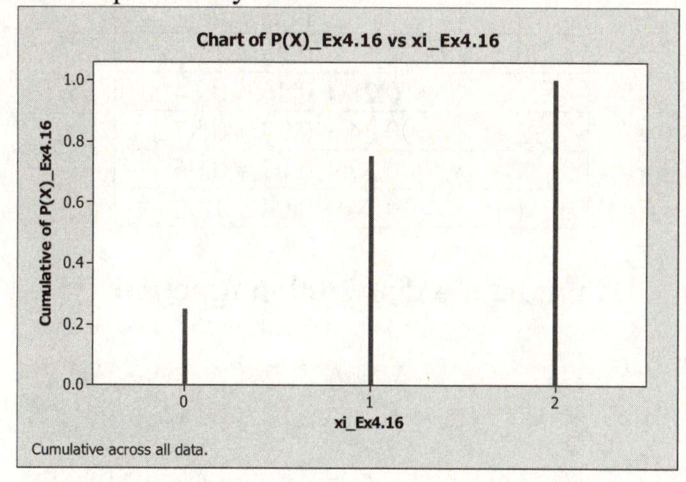

c. Find the mean

X	P(x)	XP(x)
0	.25	0
1	.50	.50
2	.25	.50
		1.00

$\mu_x = E(X) = \sum xP(x) = 1.00$

d. Find the variance of X

X	P(x)	XP(x)	(x-mu)^2	(x-mu)^2P(x)
0	.25	0	1.0	.25
1	.50	.50	0	0
2	.25	.50	1	.25
		1.00		.50

$\sigma^2_x = E[(X-\mu_x)^2] = \sum (x-\mu_x)^2 P(x) = .50$

4.18 a. Probability distribution function

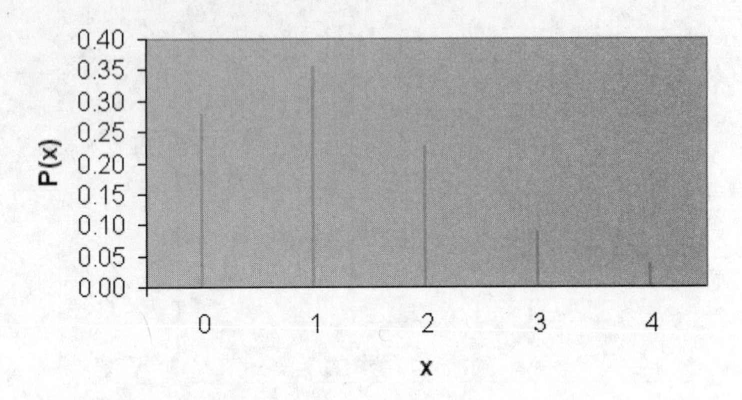

b. Cumulative probability distribution function

x	F(x)
0	0.28
1	0.28 + 0.36 = 0.64
2	0.64 + 0.23 = 0.87
3	0.87 + 0.09 = 0.96
4	0.96 + 0.04 = 1.00

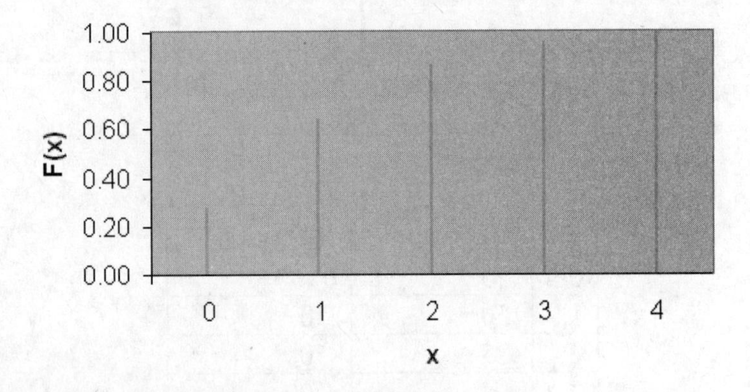

c. Find the mean of the random variable of X

x	P(x)	xP(x)
0	0.28	0
1	0.36	0.36
2	0.23	0.46
3	0.09	0.27
4	0.04	0.16
		1.25

$$\mu_x = E(X) = \sum xP(x) = 1.25$$

d. Find the variance of X

x	P(x)	xP(x)	(x-mu)^2	(x-mu)^2P(x)
0	0.28	0	1.5625	0.4375
1	0.36	0.36	0.0625	0.0225
2	0.23	0.46	0.5625	0.129375
3	0.09	0.27	3.0625	0.275625
4	0.04	0.16	7.5625	0.3025
		1.25		1.1675

$\sigma^2_x = E[(X-\mu_x)^2] = \sum (x-\mu_x)^2 P(x) = 1.1675$

4.20 a. Probability function

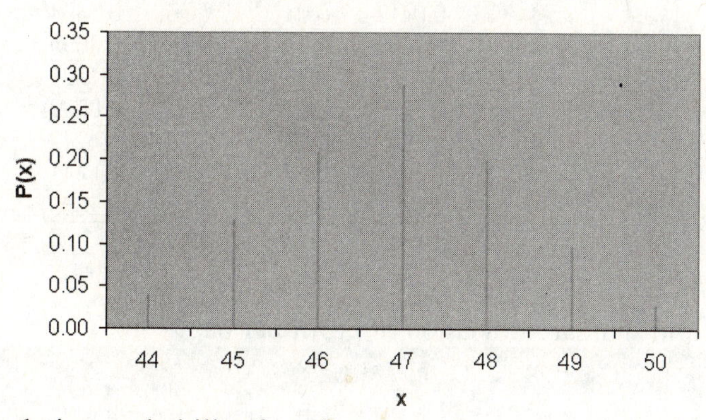

b. Cumulative probability function

x	F(x)
44	0.04
45	0.04 + 0.13 = 0.17
46	0.17 + 0.21 = 0.38
47	0.38 + 0.29 = 0.67
48	0.67 + 0.20 = 0.87
49	0.87 + 0.10 = 0.97
50	0.97 + 0.03 = 1.00

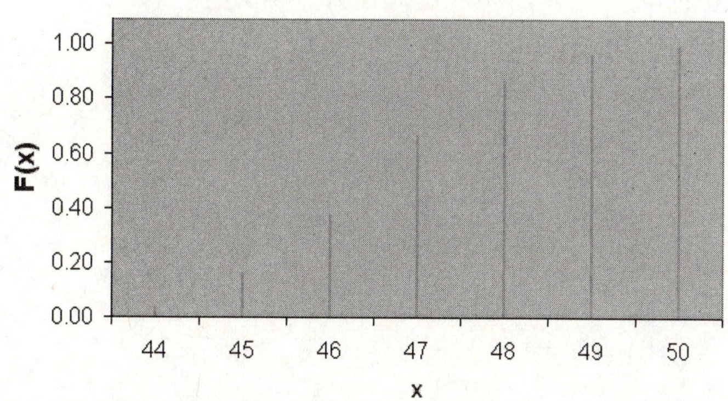

c. $P(46 \le x \le 48) = P(x=46) + P(x=47) + P(x=48) = 0.21 + 0.29 + 0.20 = 0.70$

d. P(at least one contains at least 47 lbs) = 1 – P(both packages contain fewer than 47 lbs)

$$1-[P(x<47)]^2 = 1-[0.38]^2 = 1-0.1444 = 0.8556$$

e. $\mu_x = 44(.04) + 45(.13) + 46(.21) + 47(.29) + 48(.20) + 49(.10) + 50(.03)$
= 46.9 pounds per bag

$\sigma^2_x = 1.95$ $\sigma_x = 1.3964$ pounds per bag

Excel output:

Microsoft Excel - CH 4.xls

File Edit View Insert Format Tools Data Window

N4 fx

	A	B	C	D	E
1	Weight	P(x)	F(x)	Mean	Variance
2	44	0.04	0.04	1.76	0.3364
3	45	0.13	0.17	5.85	0.4693
4	46	0.21	0.38	9.66	0.1701
5	47	0.29	0.67	13.63	0.0029
6	48	0.20	0.87	9.6	0.242
7	49	0.10	0.97	4.9	0.441
8	50	0.03	1.00	1.5	0.2883
9		1.00		46.9	1.95

f. Mean and standard deviation of profit per bag:

Microsoft Excel - CH 4.xls

File Edit View Insert Format Tools Data Window Help

Q1 fx

	J	K	L	M	N	O
1	Weight	Profit Per Bag	P(x)	F(x)	Mean	Variance
2	44	0.87	0.04	0.04	0.0348	0.00013456
3	45	0.85	0.13	0.17	0.1105	0.00018772
4	46	0.83	0.21	0.38	0.1743	0.00006804
5	47	0.81	0.29	0.67	0.2349	0.00000116
6	48	0.79	0.20	0.87	0.1580	0.00009680
7	49	0.77	0.10	0.97	0.0770	0.00017640
8	50	0.75	0.03	1.00	0.0225	0.00011532
9			1.00		0.8120	0.00078000

$\pi = 2.5 - (.75 + .02X)$

$\mu = E(\pi) = 2.5 - (.75 + (.02)(46.9)) = \$.812$

$\sigma_\pi = |.02|(1.3964) = \$.0279$

4.22 a. Probability function

X	0	1	2
P(x)	0.81	0.18	.01

$P_x(0) = (.90)(.90) = .81$

$P_x(1) = (.90)(.10) + (.10)(.90) = .18$

$P_x(2) = (.10)(.10) = .01$

b. $P(Y = 0) = 18/20 \times 17/19 = 153/190$
$P(Y=1) = (2/20 \times 18/19) + (18/20 \times 2/19) = 36/190$
$P(Y=2) = 2/20 \times 1/19 = 1/190$
The answer in part b. is different from part a. because in part b. the probability of picking a defective part on the second draw depends upon the result of the first draw.

c. $\mu = 0(.81) + .18 + 2(.01) = 0.2$ defects
$\sigma^2_x \quad = .22 - (.20)^2 = .18$

d. $\mu = 0(153/190) + (36/190) + 2(1/190) = 38/190 = 0.2$ defects
$\sigma^2_y \quad = 40/190 - (.20)^2 = .1705$

Microsoft Excel - Book1

File Edit View Insert Format Tools QIC Data Window Help

Arial

AJ6 =

	AE	AF	AG	AH	AI
1	Defects	P(x)	F(x)	Mean	Variance
2	0	0.81	0.81	0	0.03240000
3	1	0.18	0.99	0.18	0.11520000
4	2	0.01	1.00	0.02	0.03240000
5		1.00		0.2	0.18000000
6					

4.24 "One and one" $E(X) = 1(.75)(.25) + 2(.75)^2 = 1.3125$
"Two-shot foul" $E(X) = 1((.75)(.25) + (.25)(.75)) + 2(.75)^2 = 1.50$
The "two-shot foul" has a higher expected value

4.26 $\mu = 3.29 \quad \sigma^2 = 1.3259 \quad \sigma = 1.1515$

Rating	P(x)	F(x)	Mean	Variance
1	0.07	0.07	0.07	0.367087
2	0.19	0.26	0.38	0.316179
3	0.28	0.54	0.84	0.023548
4	0.30	0.84	1.20	0.15123
5	0.16	1.00	0.80	0.467856
	1.00		3.29	1.3259
			S.D.	1.151477

4.28 a. $\mu = 1.82$ breakdowns $\quad \sigma^2 = 1.0276 \quad \sigma = 1.0137$ breakdowns

Breakdowns	P(x)	F(x)	Mean	Variance	Cost	P(x)	F(x)	Mean	Variance
0	0.1	0.10	0	0.33124	0	0.1	0.10	0	745290
1	0.26	0.36	0.26	0.174824	1500	0.26	0.36	390	393354
2	0.42	0.78	0.84	0.013608	3000	0.42	0.78	1260	30618
3	0.16	0.94	0.48	0.222784	4500	0.16	0.94	720	501264
4	0.06	1.00	0.24	0.285144	6000	0.06	1.00	360	641574
	1.00		1.82	1.0276		1.00		2730	2312100
		S.D.		1.013706			S.D.		1520.559

b. Cost: C = 1500X
   E(C) = 1500(1.82) = μ = $2,730
   σ = |1500|(1.0137) = $1,520.559

4.30 Mean and variance of a Bernoulli random variable with P=.5

$$\mu_x = E(X) = \sum xP(x) = (0)(1-P) + (1)P = P = .5$$

$$\sigma^2_x = P(1-P) = .5(1-.5) = .25$$

4.32 Probability of a binomial random variable with P=.3 and n = 14, x=7 and x less than 6

Cumulative Distribution Function

Binomial with n = 14 and p = 0.3

x	P(X <= x)
0	0.006782
1	0.047476
2	0.160836
3	0.355167
4	0.584201
5	0.780516
6	0.906718
7	0.968531
8	0.991711

P(x=7) = .968531 - .906718 = .06181
P(x<6) = .7805

4.34 Probability of a binomial random variable with P=.7 and n=18, x=12 and x less than 6

Cumulative Distribution Function

Binomial with n = 18 and p = 0.7

x	P(X <= x)
0	0.000000
1	0.000000
2	0.000000
3	0.000004
4	0.000039
5	0.000269
6	0.001430
7	0.006073
8	0.020968
9	0.059586
10	0.140683
11	0.278304
12	0.465620
13	0.667345

P(x=12) = .465620 - .278304 = .1873
P(x<6) = .000269

4.36 **Cumulative Distribution Function**

Binomial with n = 5 and p = 0.250000

x	P(X <= x)
0.00	0.2373
1.00	0.6328
2.00	0.8965
3.00	0.9844
4.00	0.9990
5.00	1.0000

a. $P(x \geq 1) = 1 - Px(0) = 1 - .2373 = .7627$

b. $P(x \geq 3) = 1 - P(x \leq 2) = 1 - .8965 = .1035$

4.38 **Cumulative Distribution Function**

Binomial with n = 7 and p = 0.200000

x	P(X <= x)
0.00	0.2097
1.00	0.5767
2.00	0.8520
3.00	0.9667
4.00	0.9953
5.00	0.9996
6.00	1.0000
7.00	1.0000

$P(x \geq 2) = 1 - P(x \leq 1) = 1 - 0.5767 = 0.4233$

4.40 **Cumulative Distribution Function**

Binomial with n = 5 and p = 0.500000

x	P(X <= x)
0.00	0.0312
1.00	0.1875
2.00	0.5000
3.00	0.8125
4.00	0.9688
5.00	1.0000

a. $P(x = 5) = P(x \leq 5) - P(x \leq 4) = 1.0000 - .9688 = .0312$

b. $P(x \geq 3) = P(x \leq 5) - P(x \leq 2) = 1.0000 - .5000 = .5$

Cumulative Distribution Function

Binomial with n = 4 and p = 0.500000

x	P(X <= x)
0.00	0.0625
1.00	0.3125
2.00	0.6875
3.00	0.9375
4.00	1.0000

c. $P(x \geq 2) = 1 - P(x \leq 1) = 1 - 0.3125 = 0.6875$

d. $E(X) = np = 5(.5) = 2.5$ wins

e. $E(X) = \mu = 1 + np = 1 + 4(.5) = 3$ wins

4.42 a. **Cumulative Distribution Function**

Binomial with n = 4 and p = 0.350000

x	P(X <= x)
0.00	0.1785
1.00	0.5630
2.00	0.8735
3.00	0.9850
4.00	1.0000

$P(x \geq 2) = 1 - P(x \leq 1) = 1 - 0.5630 = 0.4370$

b. Let $Z = 2X$.

$E(Z) = 2 \cdot nP = 2 \cdot 4(0.35) = 2.8,$

$\sigma_Z = |2|\sqrt{nP(1-P)} = |2|\sqrt{4(0.35)(0.65)} = 1.908$

c. Set $E(X) = np$ equal to 2.8, where $n = 4$, and solve for P.

$4P = 2.8$, $P = 0.7$

4.44 a. $E(X) = 2000(.032) = 64$

$\sigma_x = \sqrt{2000(.032)(.968)} = 7.871$

b. Let Z = 10X

E(Z) = 10(64) = $640

$\sigma_z = |10|(7.871) = \$78.71$

4.46 a. $E(X) = \mu_x = np = 620(.78) = 483.6$, $\sigma_x = \sqrt{620(.78)(.22)} = 10.3146$

b. Let Z = 2X

E(Z) = 2(483.6) = $967.20

$\sigma_z = |2|(10.314) = \$20.6292$

4.48 The acceptance rules have the following probabilities:

(i) Rule 1: $P(X=0) = (.8)^{10} = .1074$

(ii) Rule 2: $P(X \leq 1) = (.8)^{20} + 20(.2)(.8)^{19} = .0692$

Therefore, the acceptance rule with the smaller probability of accepting a shipment containing 20% defectives will be the second acceptance rule

4.50 **Probability Density Function**

Hypergeometric with N = 50, M = 25, and n = 12

x	P(X = x)
0	0.000043
1	0.000918
2	0.008078
3	0.038706
4	0.112702
5	0.210376

P(x=5) = .210376

4.52 **Probability Density Function**

Hypergeometric with N = 80, M = 42, and n = 20

x	P(X = x)
0	0.000000
1	0.000000
2	0.000008
3	0.000093
4	0.000704
5	0.003723
6	0.014348
7	0.041322
8	0.090392
9	0.151769

$P(x=9) = .151769$

4.54 **Probability Density Function**

Hypergeometric with N = 400, M = 200, and n = 15

x	P(X = x)
0	0.000023
1	0.000375
2	0.002792
3	0.012743
4	0.039848
5	0.090434
6	0.153879
7	0.199906
8	0.199906

$P(x=8) = .1999$

4.56 **Cumulative Distribution Function**

Hypergeometric with N = 16, X = 8, and n = 8

x	P(X <= x)
1.00	0.0051
2.00	0.0660
3.00	0.3096
4.00	0.6904
5.00	0.9340
6.00	0.9949
7.00	0.9999

$P(x = 4) = P(x \leq 4) - P(x \leq 3) = .6904 - .3096 = .3808$

4.58 **Cumulative Distribution Function**

Hypergeometric with N = 10, X = 5, and n = 6

x	P(X <= x)
0.00	0.0000
1.00	0.0238
2.00	0.2619
3.00	0.7381
4.00	0.9762
5.00	1.0000

$P(x \leq 2) = .2619$

4.60 **Probability Density Function**

Poisson with mean = 2.5

x	P(X = x)
0	0.082085
1	0.205212
2	0.256516
3	0.213763
4	0.133602

P(x=4) = .1336

4.62 **Cumulative Distribution Function**

Poisson with mean = 3.5

x	P(X <= x)
0	0.030197
1	0.135888
2	0.320847
3	0.536633
4	0.725445
5	0.857614
6	0.934712

$P(x<6) = .857614$

4.64 **Cumulative Distribution Function**

Poisson with mu = 3.00000

x	P(X <= x)
0.00	0.0498
1.00	0.1991
2.00	0.4232
3.00	0.6472
4.00	0.8153
5.00	0.9161
6.00	0.9665
7.00	0.9881
8.00	0.9962
9.00	0.9989
10.00	0.9997

$P(x \leq 2) = .4232$

4.66 **Cumulative Distribution Function**

Poisson with mu = 4.20000

x	P(X <= x)
0.00	0.0150
1.00	0.0780
2.00	0.2102
3.00	0.3954
4.00	0.5898
5.00	0.7531
6.00	0.8675
7.00	0.9361
8.00	0.9721
9.00	0.9889
10.00	0.9959

$P(x \geq 3) = 1 - P(x \leq 2) = 1 - .2102 = .7898$

4.68 **Cumulative Distribution Function**

Poisson with mu = 5.50000

x	P(X <= x)
0.00	0.0041
1.00	0.0266
2.00	0.0884
3.00	0.2017
4.00	0.3575
5.00	0.5289
6.00	0.6860
7.00	0.8095
8.00	0.8944
9.00	0.9462
10.00	0.9747

$P(x \leq 2) = .0884$

4.70 **Cumulative Distribution Function**

Poisson with mu = 6.00000

x	P(X <= x)
0.00	0.0025
1.00	0.0174
2.00	0.0620
3.00	0.1512
4.00	0.2851
5.00	0.4457
6.00	0.6063
7.00	0.7440
8.00	0.8472
9.00	0.9161
10.00	0.9574

$P(x \geq 3) = 1 - P(x \leq 2) = 1 - .0620 = .9380$

4.72 Two models are possible – the poisson distribution is appropriate when the warehouse is serviced by many thousands of independent truckers where the mean number of 'successes' is relatively small. However, under the assumption of a small fleet of 10 trucks with a probability of any truck arriving during a given hour is .1, then the binomial distribution is the more appropriate model. Both models yield similar, although not identical, probabilities.

Cumulative Distribution Function

Poisson with mean = 1

x	P(X <= x)
0	0.36788
1	0.73576
2	0.91970
3	0.98101
4	0.99634
5	0.99941
6	0.99992
7	0.99999
8	1.00000
9	1.00000
10	1.00000

Cumulative Distribution Function

Binomial with n = 10 and p = 0.1

x	P(X <= x)
0	0.34868
1	0.73610
2	0.92981
3	0.98720
4	0.99837
5	0.99985
6	0.99999
7	1.00000
8	1.00000
9	1.00000
10	1.000

4.74 a. Compute marginal probability distributions for X and Y

Exercise_4.74			X_4.74				
Y_4.74		1	2	P(y)	Mean of Y	Var of Y	StDev of Y
	0	0.2	0.25	0.45	0	0.136125	
	1	0.3	0.25	0.55	0.55	0.111375	
P(x)		0.5	0.5	1	0.55	0.2475	0.497494
Mean of X		0.5	1	1.5			
Var of X		0.125	0.125	0.25			
StDev of X				0.5			
xyP(x)		0.3	0.5	0.8			
Cov(x,y) =							
sum xyP(x)-muxmuy		-0.025					

b. Compute the covariance and correlation for X and Y

$$Cov(X,Y) = \sum_x \sum_y xyP(x,y) - \mu_x\mu_y = .80 - (1.5)(.55) = -.025$$

$$\rho = Corr(X,Y) = \frac{Cov(X,Y)}{\sigma_x\sigma_y} = -.025/(.5)(.497494) = -.1005$$

4.76 a. Compute marginal probability distributions for X and Y

Exercise_4.76			X_4.76				
Y_4.76		1	2	P(y)	Mean of Y	Var of Y	StDev of Y
	0	0.3	0.2	0.5	0	0.125	
	1	0.25	0.25	0.5	0.5	0.125	
P(x)		0.55	0.45	1	0.5	0.25	0.5
Mean of X		0.55	0.9	1.45			
Var of X		0.55	1.8	2.35			
StDev of X				1.532971			
xyP(x)		0.25	0.5	0.75			
Cov(x,y) =							
sum xyP(x)-muxmuy		0.025					

b. Compute the covariance and correlation for X and Y

$$Cov(X,Y) = \sum_x \sum_y xyP(x,y) - \mu_x\mu_y = .75 - (1.45)(.5) = 0.025$$

$$\rho = Corr(X,Y) = \frac{Cov(X,Y)}{\sigma_x\sigma_y} = 0.025/(1.53297)(.5) = 0.0326$$

c. Compute the mean and variance for the linear function W = 2X + Y

$$\mu_W = a\mu_x + b\mu_y = (2)1.45 + (1).5 = 3.4$$

$$\sigma^2_W = a^2\sigma^2_X + b^2\sigma^2_Y + 2abCov(X,Y) = 2^2(2.35) + 1^2(.25) + 2(2)(1)(0.025) = 9.75$$

4.78 a. Compute the marginal probability distributions for X and Y

		1	2	P(y)	Mean of Y	Var of Y	StDev of Y
	0	0.25	0.25	0.5	0	0.125	
	1	0.25	0.25	0.5	0.5	0.125	
P(x)		0.5	0.5	1	0.5	0.25	0.5
Mean of X		0.5	1	1.5			
Var of X		0.125	0.125	0.25			
StDev of X				0.5			
xyP(x)		0.25	0.5	0.75			
Cov(x,y) =							
sum xyP(x)-muxmuy		0					

b. Compute the covariance and correlation for X and Y

$$Cov(X,Y) = \sum_x \sum_y xyP(x,y) - \mu_x\mu_y = .75 - (1.5)(.5) = 0.0$$

$$\rho = Corr(X,Y) = \frac{Cov(X,Y)}{\sigma_x\sigma_y} = 0.0/(.5)(.5) = 0.0$$

Note that when covariance between X and Y is equal to zero, it follows that the correlation between X and Y is also zero.

c. Compute the mean and variance for the linear function W = X - Y

$$\mu_W = a\mu_x + b\mu_y = (1)1.5 + (-1).5 = 1.0$$

$$\sigma^2_W = a^2\sigma^2_X + b^2\sigma^2_Y + 2abCov(X,Y) = 1^2(.25) + 1^2(.25) + 2(1)(-1)(0.0) = .5$$

4.80 a. ompute the marginal probability distributions for X and Y.

Exercise_4.80			X_4.80				
Y_4.80		1	2	P(y)	Mean of Y	Var of Y	StDev of Y
	0	0	0.6	0.6	0	0.096	
	1	0.4	0	0.4	0.4	0.144	
P(x)		0.4	0.6	1	0.4	0.24	0.489898
Mean of X		0.4	1.2	1.6			
Var of X		0.144	0.096	0.24			
StDev of X				0.489898			
xyP(x)		0.4	0	0.4			
Cov(x,y) =							
sum xyP(x)-muxmuy		-0.24					

b. Compute the covariance and correlation for X and Y

$$Cov(X,Y) = \sum_x \sum_y xyP(x,y) - \mu_x\mu_y = .40 - (1.6)(.4) = -0.24$$

$$\rho = Corr(X,Y) = \frac{Cov(X,Y)}{\sigma_x\sigma_y} = -0.24/(.489898)(.489898) = -1.00$$

c. Compute the mean and variance for the linear function W = 2X - 4Y

$\mu_W = a\mu_x + b\mu_y = (2)1.6 + (-4).4 = 1.6$

$\sigma^2_W = a^2\sigma^2_X + b^2\sigma^2_Y + 2abCov(X,Y) = 2^2(.24) + (-4)^2(.24) + 2(2)(-4)(-.24) = 8.64$

4.82 a. Px(0) = .07 + .07 + .06 + .02 = .22
Px(1) = .09 + .06 + .07 + .04 = .26
Px(2) = .06 + .07 + .14 + .16 = .43
Px(3) = .01 + .01 + .03 + .04 = .09
$\mu_x = 0 + .26 + 2(.43) + 3(.09) = 1.39$

b. Py(0) = .07 + .09 + .06 + .01 = .23
Py(1) = .07 + .06 + .07 + .01 = .21
Py(2) = .06 + .07 + .14 + .03 = .30
Py(3) = .02 + .04 + .16 + .40 = .26
$\mu_y = 0 + .21 + 2(.3) + 3(.26) = 1.59$

c. $P_{Y|X}(0|3) = .01/.09 = .1111$
$P_{Y|X}(1|3) = .01/.09 = .1111$
$P_{Y|X}(2|3) = .03/.09 = .3333$
$P_{Y|X}(3|3) = .04/.09 = .4444$

d. $Cov(X,Y) = E(XY) - \mu_x\mu_y$
E(XY) = 0 + 1(1)(.06) + 1(2)(.07) + 1(3)(.04) + 2(1)(.07) + 2(2)(.14) + 2(3)(.16) + 3(1)(.01) + 3(2)(.03) + 3(3)(.04) = 2.55
$Cov(X,Y) = 2.55 - (1.39)(1.59) = .3399$

e. No, because $Cov(X,Y) \neq 0$

4.84 a. Py(0) = .08 + .03 + .01 = .12
Py(1) = .13 + .08 + .03 = .24
Py(2) = .09 + .08 + .06 = .23
Py(3) = .06 + .09 + .08 = .23
Py(4) = .03 + .07 + .08 = .18

b. $P_{Y|X}(y|3)$ = 1/26; 3/26; 6/26; 8/26; 8/26

c. No, because Px,y(3,4) = .08 $\neq$ .0468 = Px(3)Py(4)

4.86 a.

Y/X	0	1	Total
0	.704	.168	.872
1	.096	.032	.128
Total	.80	.20	1.00

b. $P_{Y|X}(y|0)$ = .88; .12

c. Px(0) = .80
Px(1) = .20
Py(0) = .872
Py(1) = .128

d. E(XY) = .032;
$\mu_x = 0 + 1(.20) = .20$, $\mu_y = 0 + 1(.128) = .128$
$Cov(X,Y) = .032 - (.20)(.128) = .0064$
The covariance indicates that there is a positive association between X and Y, professors are more likely to be away from the office on Friday than during the other days.

4.88 See table above. Number of total complaints (food complaints + service complaints) has a mean of (1.36 + 1.64) = 3.00. If the two types of complaints are independent, then the variance of total complaints is equal to the sum of the variance of the two types of complaints because the covariance would be zero. (.8104 + .7904) = 1.6008. The standard deviation will be the square root of the variance = 1.26523.
If the number of food and service complaints are not independent of each other, then the covariance would no longer be zero. The mean would remain the same; however, the standard deviation would change. The variance of the sum of the two types of complaints becomes the variance of one plus the variance of the other plus two times the covariance.

4.90 a. No, not necessarily. There is a probability distribution associated with the rates of return in the mutual fund and not all rates of return will equal the expected value.
b. Which fund to invest in will depend not only on the expected value of the return but also on the riskiness of each fund and how risk averse the client is.

4.92

Cars	P(x)	F(x)	Mean	Variance
0	0.1	0.10	0	0.48841
1	0.2	0.30	0.2	0.29282
2	0.35	0.65	0.7	0.015435
3	0.16	0.81	0.48	0.099856
4	0.12	0.93	0.48	0.384492
5	0.07	1.00	0.35	0.544887
Ex 4.92	1.00		2.21	1.8259
		S.D.		1.351259

a. E(X) = 2.21 cars sold
b. Standard deviation = 1.3513 cars
c. Mean Salary = \$250 + \$300 (2.21) = \$913. Standard deviation of salary = \$300(1.3513) = \$405.39
d. To earn a salary of \$1,000 or more, the salesperson must sell at least 3 cars. $P(X \geq 3) = .16 + .12 + .07 = .35$

4.94 a. Positive covariance: Consumption expenditures & Disposable income
b. Negative covariance: Price of cars and the number of cars sold
c. Zero covariance: Dow Jones stock market average & rainfall in Brazil

4.96

			X Years					
Y Visits	1	2	3	4	P(y)	Mean of Y	Var of Y	StDev of Y
0	0.07	0.05	0.03	0.02	0.17	0	0.2057	
1	0.13	0.11	0.17	0.15	0.56	0.56	0.0056	
2	0.04	0.04	0.09	0.1	0.27	0.54	0.2187	
P(x)	0.24	0.2	0.29	0.27	1	1.1	0.43	0.6557439
Mean of X	0.24	0.4	0.87	1.08	2.59			
Var of X	0.606744	0.0696	0.048749	0.5368	1.2619			
StDev of X					1.12334			
xyP(x)	0.21	0.38	1.05	1.4	3.04			
sum xyP(x)*muxmuy	0.191							

a. Py(0) = .07+.05+.03+.02 = .17

b. E(X) = $\mu_x = .24 + 2(.2) + 3(.29) + 4(.27) = 2.59$

E(Y) = $\mu_y = .56 + 2(.27) = 1.1$

c. E(XY)=3.04, Cov(X,Y) = 3.04 – (2.59)(1.1) = .191. This implies that there is a positive relationship between the number of years in school and the number of visits to a museum in the last year.

4.98 a. $P(x=3) = \binom{5}{3} .55^3 .45^2 = .3369$

b. $P(x \geq 3) = P(3)+P(4)+P(5) = .3369 + (5)(.55)^4(.45) + (1)(.55)^5(1) =$ .5931

c. $\mu = np = (80)(.55) = 44$ will graduate in 4 years. The proportion is 44/80 = .55. $\sigma = \sqrt{80(.55)(.45)} = 4.4497$. The proportion is 4.4497/80 = .05562

4.100 To evaluate the effectiveness of the analyst's ability, find the probability that x is greater than or equal to 3 at random.

$$P(x \geq 3) = \frac{\binom{5}{3}\binom{10}{2}}{\binom{15}{5}} + \frac{\binom{5}{4}\binom{10}{1}}{\binom{15}{5}} + \frac{\binom{5}{5}\binom{10}{0}}{\binom{15}{5}} = .16683$$

4.102 a. $P(0) = e^{-2.4} = .09072$

b. $P(x > 3) = 1 - e^{-2.4} - e^{-2.4}(2.4) - e^{-2.4}(2.4)^2/2! - e^{-2.4}(2.4)^3/3! = .2213$

4.104 $P(x=0) = e^{-2.4} = .0907$
Let Y be the number of stalls for both lines.
Find the $P(Y \geq 1) = 1 - P(Y=0) = 1 - (.0907)^2 = .99177$

4.106 Compute the mean and variance.

Exercise_4.106			X_4.106					
Y_4.106		3	4	5	P(y)	Mean of Y	Var of Y	StDev of Y
	4	0.1	0.15	0.05	0.3	1.2	1.2	
	6	0.1	0.2	0.1	0.4	2.4	3.15544E-31	
	8	0.05	0.15	0.1	0.3	2.4	1.2	
P(x)		0.25	0.5	0.25	1	6	2.4	1.549193
Mean of X		0.75	2	1.25	4			
Var of X		0.25	0	0.25	0.5			
StDev of X					0.707107			
xyP(x)		4.2	12	8	24.2			
cov(x,y) =								
sum xyP(x)-muxmuy		0.2						

Compute the mean and variance for the linear function W = aX – bY.

$\mu_W = a\mu_x + b\mu_y = (10)4 - (5)6 = 10$

$\sigma_W^2 = a^2\sigma_x^2 + b^2\sigma_y^2 + 2abCov(x,y) = 10^2(0.5) + 5^2(2.4) + 2(10)(5)(0.2) = 130$

Therefore, the mean of the trade balance is \$10,000. The variance of the trade balance is \$130,000.

4.108

a. Shown below is the cumulative distribution function for a Poisson distribution with $\lambda = 19.5$.

Cumulative Distribution Function

Poisson with mu = 19.5000

x	P(X <= x)
0.00	0.0000
1.00	0.0000
2.00	0.0000
3.00	0.0000
4.00	0.0000
5.00	0.0001
6.00	0.0004
7.00	0.0011
8.00	0.0028
9.00	0.0067
10.00	0.0141
11.00	0.0273
12.00	0.0488

13.00	0.0809
14.00	0.1257
15.00	0.1840
16.00	0.2550
17.00	0.3364
18.00	0.4246
19.00	0.5151
20.00	0.6034
21.00	0.6854
22.00	0.7580
23.00	0.8196
24.00	0.8697
25.00	0.9087
26.00	0.9380
27.00	0.9591
28.00	0.9739
29.00	0.9838
30.00	0.9902

The cumulative distribution function shows that there is a 90.87% chance that there are 25 applications or fewer. Thus, assuming a five-day work week, 25/5 = 5 analysts should be hired.

b. P(2 of the 5 analysts have no clients) = P(3 of 5 have at least 1 client)
In this case, there will be a minimum of 11 clients (5 for the first 2 analysts and 1 for the third analyst) and a maximum of 15 clients (5 for each of the first 3 analysts).
$P(11 \le x \le 15) = P(x \le 15) - P(x \le 10) = 0.1840 - 0.0141 = 0.1699$

c. The probability that customers would cancel given that 4 analysts are hired is the probability that there are more applications than the analysts could complete $(4 \cdot 5 = 20)$.
$P(x > 20) = 1 - P(x \le 20) = 1 - 0.6034 = 0.3966$

d. P(2 of the 4 analysts have no clients) = P(2 of 4 have at least 1 client)
In this case, there will be a minimum of 6 clients (5 for the first analyst and 1 for the second analyst) and a maximum of 10 clients (5 for each of the first 2 analysts).
$P(5 \le x \le 10) = P(x \le 10) - P(x \le 4) = 0.0141 - 0.0000 = 0.0141$

Chapter 5:

Continuous Random Variables and Probability Distributions

5.2 $P(1.0 < X < 1.9) = F(1.9) - F(1.0) = (.5)(1.9) - (.5)(1.0) = 0.45$

5.4 $P(X > 1.3) = F(1.3) = (.5)(2.0) - (.5)(1.3) = 0.35$

5.6 a.

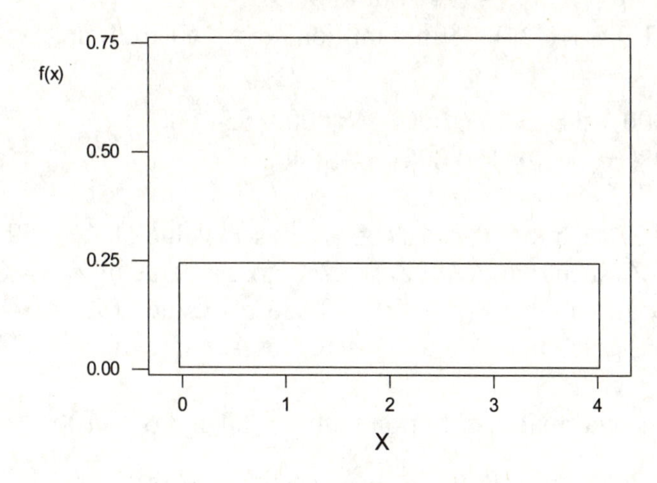

b.

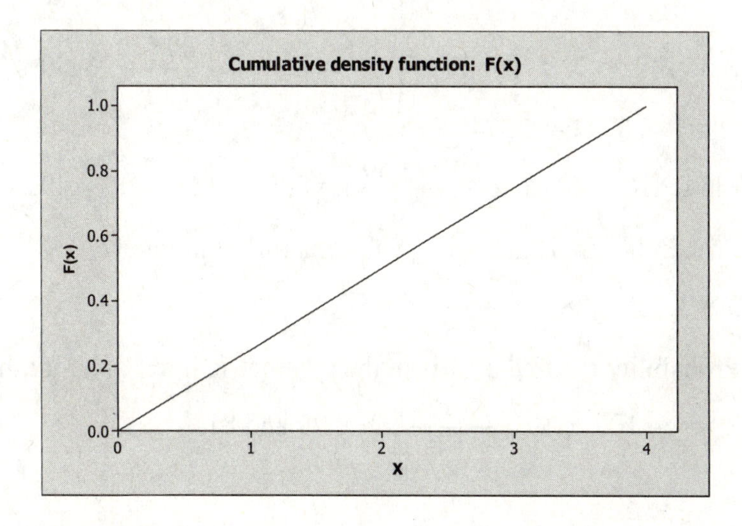

c. $P(x < 1) = .25$

d. $P(X < .5) + P(X > 3.5) = P(X < .5) + 1 - P(X < 3.5) = .25$

5.8 a. $P(380 < X < 460) = P(X < 460) - P(X < 380) = .6 - .4 = .2$
b. $P(X < 380) < (PX< 400) < P(X < 460)$; $.4 < P(X < 400) < .6$

5.10 W = a + bX. If Available Funds = 1000 - 2X where X = number of units produced, find the mean and variance of the profit if the mean and variance for the number of units produced are 50 and 90 respectively. $\mu_W = a + b\mu_x =$ $1000 - 2(50) = 900$. $\sigma^2_W = b^2\sigma^2_X = (-2)^2(90) = 360$.

5.12 W = a + bX. If Available Funds = 6000 – 3X where X = number of units produced, find the mean and variance of the profit if the mean and variance for the number of units produced are 1000 and 900 respectively.
$\mu_W = a + b\mu_x = 6000 - 2(1000) = 4000$. $\sigma^2_W = b^2\sigma^2_X = (-3)^2(900) = 8100$

5.14 $\mu_Y = 20 + \mu_X = 20 + 4 = \$24$ million
Bid = 1.1 μ_Y =1.1(24) = \$26.4 million, σ_π = \$1 million

5.16 $\mu_Y = 6{,}000 + .08\,\mu_X = 6{,}000 + 48{,}000 = \$54{,}000$
$\sigma_Y = |.08|\,\sigma_X = .08(180{,}000) = \$14{,}400$

5.18 a. Find Z_0 such that $P(Z < Z_0) = .7$, closest value of $Z_0 = .52$
b. Find Z_0 such that $P(Z < Z_0) = .25$, closest value of $Z_0 = -.67$
c. Find Z_0 such that $P(Z > Z_0) = .2$, closest value of $Z_0 = .84$
d. Find Z_0 such that $P(Z > Z_0) = .6$, closest value of $Z_0 = -.25$

5.20 X follows a normal distribution with $\mu = 80$ and $\sigma^2 = 100$
a. Find P(X > 60). $P(Z > \frac{60-80}{10}) = P(Z > -2.00) = .5 + .4772 = .9772$
b. Find P(72 < X < 82). $P(\frac{72-80}{10} < Z < \frac{82-80}{10}) = P(-.80 < Z < .20) =$ $.2881 + .0793 = .3674$
c. Find P(X < 55). $P(Z < \frac{55-80}{10}) = P(Z < -2.50) = .5 - .4938 = .0062$
d. Probability is .1 that X is greater than what number?
$Z = 1.28$. $1.28 = \frac{X-80}{10}$ $X = 92.8$
e. Probability is .08 that X is in the symmetric interval about the mean between?
$Z = +/- .10$. $\pm.10 = \frac{X-80}{10}$. $X = 79$ and 81.

5.22 a. $P(Z < \frac{400-380}{50}) = P(Z < .4) = .6554$

b. $P(Z > \frac{360-380}{50}) = P(Z > -.4) = F_Z(.4) = .6554$

c. The graph should show the property of symmetry – the area in the tails equidistant from the mean will be equal.

d. $P(\frac{300-380}{50} < Z < \frac{400-380}{50}) = P(-1.6 < Z < .4) = F_Z(.4) - [1 - F_Z(1.6)] =$ $.6554 - .0548 = .6006$

e. The area under the normal curve is equal to .8 for an infinite number of ranges – merely start at a point that is marginally higher. The shortest range will be the one that is centered on the z of zero. The z that corresponds to an area of .8 centered on the mean is a Z of ±1.28. This yields an interval of the mean plus and minus $64: [$316, $444]

5.24 a. $P(Z > \frac{38-35}{4}) = P(Z > .75) = 1 - F_Z(.75) = .2266$

b. $P(Z < \frac{32-35}{4}) = P(Z < -.75) = 1 - F_Z(.75) = .2266$

c. $P(\frac{32-35}{4} < Z < \frac{38-35}{4}) = P(-.75 < Z < .75) = 2F_Z(.75) - 1 = 2(.7734) -$ $1 = .5468$

d. (i) The graph should show the property of symmetry – the area in the tails equidistant from the mean will be equal.
(ii) The answers to a, b, c sum to one because the events cover the entire area under the normal curve which by definition, must sum to 1.

5.26 a. $P(Z < \frac{10-12.2}{2.8}) = P(Z < -.79) = 1 - F_z(.79) = .2148$

b. $P(Z > \frac{15-12.2}{2.8}) = P(Z > 1) = 1 - F_z(1) = .1587$

c. $P(\frac{12-12.2}{2.8} < Z < \frac{15-12.2}{2.8}) = P(-.07 < Z < 1) = F_z(1) - [1 - F_z(.07)] =$ $.8413 - .4721 = .3692$

d. The answer to a. will be larger because 10 grams is closer to the mean than is 15 grams. Thus, there would be a greater area remaining less than 10 grams than will be the area above 15 grams.

5.28 $P(Z > 1.5) = 1 - F_z(1.5) = .0668$

5.30 $P(Z > .67) = .25,\ .67\sigma = 17.8 - \mu$
$P(Z > 1.03) = .15,\ 1.04\sigma = 19.2 - \mu$
Solving for μ, σ: $\mu = 15.265, \sigma^2 = (3.7838)^2 = 14.317$

5.32 For Investment A, the probability of a return higher than 10%:

$P(Z > \frac{10-10.4}{1.2}) = P(Z > -.33) = F_Z(.33) = .6293$

For Investment B, the probability of a return higher than 10%

$P(Z > \frac{10-11.0}{4}) = P(Z > -.25) = F_Z(.25) = .5987$

Therefore, Investment A is a better choice

5.34 a. $P(Z > -1.28) = .9$, $-1.28 = \frac{Xi-150}{40}$, $Xi = 98.8$

b. $P(Z < .84) = .8$, $.84 = \frac{Xi-150}{40}$, $Xi = 183.6$

c. $P(X \geq 1) = 1 - P(X = 0) = 1-[P(Z< \frac{120-150}{40})]^2 = 1 - [P(Z < -.75)]^2 =$
$1 - (.2266)^2 = .9487$

5.36 a. $P(\frac{400-420}{80} < Z < \frac{480-420}{80}) = P(-.25 < Z < .75) =$
$F_z(.75) - [1 - F_Z(.25)] = .7734 - .4013 = .3721$

b. $P(Z > 1.28) = .1$, $1.28 = \frac{Xi-420}{80}$, $Xi = 522.4$

c. 400 – 439

d. 520 – 559

e. $P(X \geq 1) = 1 - P(X = 0) = 1 - [P(Z< \frac{500-420}{80})]^2 = 1 - (.8413)^2 = .2922$

5.38 $P(Z < 1.5) = .9332$, $1.5 = \frac{85-70}{\sigma}$, $\sigma = 10$

$P(Z > \frac{80-70}{10}) = P(Z > 1) = .1587$

$P(X \geq 1) = 1 - P(X=0) = 1 - [F_Z(1)]^4 = 1 - (.8413)^4 = .4990$

5.40 n = 1600 from a binomial probability distribution with P = .40

a. Find $P(X > 1650)$. $E[X] = \mu = 1600(.4) = 640$,

$\sigma = \sqrt{(1600)(.4)(.6)} = 19.5959$ $P(Z > \frac{1650-1600}{19.5959})$

$= P(Z > 2.55) = 1 - F_Z(2.55) = .0054$

b. Find $P(X < 1530)$.

$P(Z < \frac{1530-1600}{19.5959}) = P(Z < -3.57) = 1 - F_Z(3.57) = .0002$

c. $P(\frac{1550-1600}{19.5959} < Z < \frac{1650-1600}{19.5959})$
$= P(-2.55 < Z < 2.55) = (2)F_z(2.55) = (2).4946 = .9892$

d. Probability is .09 that the number of successes is less than how many?
$Z = -1.34$.
$-1.34 = \frac{X-1600}{19.5959}$ $X = 1573.741$
$\approx$1,574 successes

e. Probability is .20 the number of successes is greater than? $Z = .84$.
$.84 = \frac{X-1600}{19.5959}$. $X = 1616.46 \approx$1,616 successes

5.42 n = 1600 from a binomial probability distribution with P = .40

a. Find P(P > .45). $E[P] = \mu = P = .40$, $\sigma = \sqrt{\frac{P(1-P)}{n}} = \sqrt{\frac{.4(1-.4)}{1600}}$
$= .01225$ $P(Z > \frac{.45-.40}{.01225}) = P(Z > 4.082) = 1 - F_Z(4.082) = .0000$

b. Find P(P < .36). $P(Z < \frac{.36-.40}{.01225}) = P(Z < -3.27) = 1 - F_Z(3.27) = .0005$

c. $P(\frac{.44-.40}{.01225} < Z < \frac{.37-.40}{.01225}) = P(3.27 < Z < -2.45) = 1 - [(2)[1-F_z(3.27)]]$
$= 1 - (2)[1-.9995] = .9995 - .0071 = .9924$

d. Probability is .20 that the percentage of successes is less than what percent? $Z = -.84$. $-.84 = \frac{X-.40}{.01225}$ $P = 38.971\%$

e. Probability is .09 the percentage of successes is greater than?
$Z = 1.34$. $1.34 = \frac{X-.40}{.01225}$. $P = 41.642\%$

5.44 a. $E[X] = \mu = 900(.2) = 180$, $\sigma = \sqrt{(900)(.2)(.8)} = 12$

$P(Z > \frac{200-180}{12}) = P(Z > 1.67) = 1 - F_Z(1.67) = .0475$

b. $P(Z < \frac{175-180}{12}) = P(Z < -.42) = 1 - F_Z(.42) = .3372$

5.46 $E[X] = (100)(.6) = 60$, $\sigma = \sqrt{(100)(.6)(.4)} = 4.899$

$P(Z < \frac{50-60}{4.899}) = P(Z < -2.04) = 1 - F_Z(2.04) = 1 - .9793 = .0207$

5.48 $P(Z > \frac{38-35}{4}) = P(Z > .75) = 1 - F_Z(.75) = 1 - .7734 = .2266$

$E[X] = 100(.2266) = 22.66, \sigma = \sqrt{(100)(.2266)(.7734)} = 4.1863$

$P(Z > \frac{25-22.66}{4.1863}) = P(Z > .56) = 1 - F_Z(.56) = 1 - .7123 = .2877$

5.50 $\lambda = 1.0$, what is the probability that an arrival occurs in the first t=2 time units?

Cumulative Distribution Function

```
Exponential with mean = 1
x   P( X <= x )
0      0.000000
1      0.632121
2      0.864665
3      0.950213
4      0.981684
5      0.993262
```

$P(T < 2) = .864665$

5.52 $\lambda = 5.0$, what is the probability that an arrival occurs after t=7 time units?

Cumulative Distribution Function

```
Exponential with mean = 5
x   P( X <= x )
0      0.000000
1      0.181269
2      0.329680
3      0.451188
4      0.550671
5      0.632121
6      0.698806
7      0.753403
8      0.798103
```

$P(T>7) = 1-[P(T \le 8)] = 1 - .7981 = .2019$

5.54 $\lambda = 3.0$, what is the probability that an arrival occurs after t=2 time units?

Cumulative Distribution Function

```
Exponential with mean = 3
x   P( X <= x )
0      0.000000
1      0.283469
2      0.486583
3      0.632121
```

$P(T<2) = .4866$

5.56 $P(X > 18) = e^{-(18/15)} = .3012$

5.58 a. $P(X > 3) = 1 - [1 - e^{-(3/\mu)}] = e^{-3\lambda}$ since $\lambda = 1/\mu$

b. $P(X > 6) = 1 - [1 - e^{-(6/\mu)}] = e^{-(6/\mu)} = e^{-6\lambda}$

c. $P(X>6|X>3) = P(X > 6)/P(X > 3) = e^{-6\lambda} / e^{-3\lambda}] = e^{-3\lambda}$

The probability of an occurrence within a specified time in the future is not related to how much time has passed since the most recent occurrence.

5.60 Let $\lambda = 20 \text{ trucks}/60 \text{ minutes} = 1 \text{ truck}/3 \text{ minutes}$.

a. $P(t \geq 5) = 1 - P(t < 5) = 1 - [1 - e^{-(1/3)(5)}] = 0.1889$

b. $P(t \leq 1) = 1 - e^{-(1/3)(2)} = 0.4866$

c. $P(4 \leq t \leq 10) = [1 - e^{-(1/3)(10)}] - [1 - e^{-(1/3)(4)}] = 0.2279$

5.62 Find the mean and variance of the random variable: W = 5X + 4Y with correlation = -.5

$\mu_W = a\mu_x + b\mu_y$ = 5(100) + 4(200) = 1300

$\sigma^2{}_W = a^2\sigma^2{}_X + b^2\sigma^2{}_Y + 2abCorr(X,Y)\sigma_X\sigma_Y$

$= 5^2(100) + 4^2(400) + 2(5)(4)(-.5)(10)(20) = 4{,}900$

5.64 Find the mean and variance of the random variable: W = 5X – 4Y with correlation = .5.

$\mu_W = a\mu_x - b\mu_y$ = 5(500) – 4(200) = 1700

$\sigma^2{}_W = a^2\sigma^2{}_X + b^2\sigma^2{}_Y - 2abCorr(X,Y)\sigma_X\sigma_Y$

$= 5^2(100) + 4^2(400) - 2(5)(4)(.5)(10)(20) = 4900$

5.66 μ_Z = 100,000(.1) + 100,000(.18). μ_X = 10,000 + 18,000 = 28,000

σ_Z = 0. Note that the first investment yields a certain profit of 10% which is a zero standard deviation. σ_X = 100,000(.06) = 6,000

5.68 $\mu_Z = \mu_1 + \mu_2 + \mu_3$ = 50,000 + 72,000 + 40,000 = 162,000

$\sigma_Z = \sqrt{\sigma_1^2 + \sigma_2^2 + \sigma_3^2} = \sqrt{(10{,}000)^2 + (12{,}000)^2 + (9{,}000)^2} = 18{,}027.76$

5.70 The calculation of the mean is correct, but the standard deviations of two random variables cannot be summed. To get the correct standard deviation, add the variances together and then take the square root. The standard deviation: $\sigma = \sqrt{5(16)^2} = 35.7771$

5.72 a. Compute the mean and variance of the portfolio with correlation of +.5

$\mu_W = a\mu_x + b\mu_y$ = 50(25) + 40(40) = 2850

$\sigma^2{}_W = a^2\sigma^2{}_X + b^2\sigma^2{}_Y + 2abCorr(X,Y)\sigma_X\sigma_Y$

$= 50^2(121) + 40^2(225) + 2(50)(40)(.5)(11)(15) = 992{,}500$

b. Recompute with correlation of -.5

$\mu_W = a\mu_x + b\mu_y$ = 50(25) + 40(40) = 2850

$\sigma^2{}_W = a^2\sigma^2{}_X + b^2\sigma^2{}_Y + 2abCorr(X,Y)\sigma_X\sigma_Y$

$= 50^2(121) + 40^2(225) + 2(50)(40)(-.5)(11)(15) = 332{,}500$

5.74 a. $W = aX - bY = 10X - 10Y$

$\mu_W = a\mu_x - b\mu_y$ = 10(100) – 10(90) = 100

$\sigma^2_W = a^2\sigma^2_X + b^2\sigma^2_Y - 2abCorr(X,Y)\sigma_X\sigma_Y$

$=10^2(100) + 10^2(400) - 2(10)(10)(-.4)(10)(20) = 66{,}000$

$\sigma_W = \sqrt{66{,}000}$ =256.90465

b. $P(Z < \frac{0-100}{256.90465}) = P(Z < -.39) = 1 - F_Z(.39) = 1 - .6517 = .3483$

5.76 a. $W = aX - bY = 1X - 1Y$

$\mu_W = a\mu_x - b\mu_y$ = 1(100) – 1(105) = -5

$\sigma^2_W = a^2\sigma^2_X + b^2\sigma^2_Y - 2abCorr(X,Y)\sigma_X\sigma_Y$

$=1^2(900) + 1^2(625) - 2(1)(1)(.7)(30)(25) = 475$ $\sigma_W = \sqrt{475}$ =21.79449

b. $P(Z > \frac{0-(-5)}{21.79449}) = P(Z > .23) = 1 - F_Z(.23) = 1 - .5910 = .4090$

5.78 a.

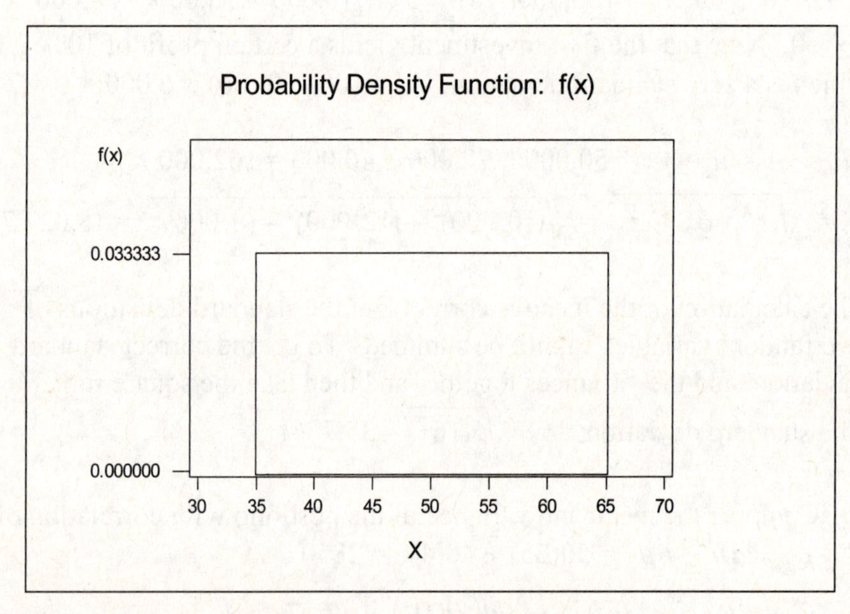

b. Cumulative density function

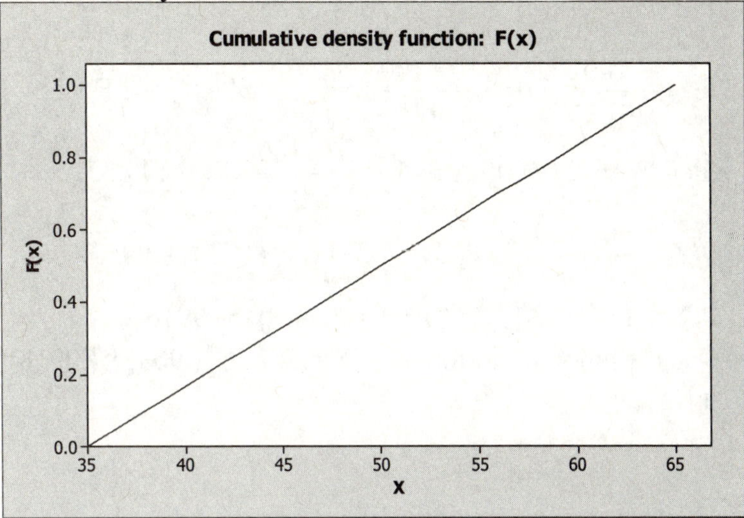

c. $P(40 < X < 50) = (50/30) - (40/30) = 10/30$

d. $E[X] = \dfrac{65+35}{2} = 50$

5.80 a. $\mu_Y = 2000(1.1) + 1000(1 + \mu_x) = 2{,}200 + 1{,}160 = 3{,}360$

b. $\sigma_Y = |1000|\ \sigma_X = 1000(.08) = 80$

5.82 Given that the variance of both predicted earnings and forecast error are both positive and given that the variance of actual earnings is equal to the sum of the variances of predicted earnings and forecast error, then the Variance of predicted earnings must be less than the variance of actual earnings

5.84 a. $P(Z > \dfrac{3-2.6}{.5}) = P(Z > .8) = 1 - F_Z(.8) = .2119$

b. $P(\dfrac{2.25-2.6}{.5} < Z < \dfrac{2.75-2.6}{.5}) = P(-.7 < Z < .3) = F_z(.3) - [1 - F_Z(.3)] =$ .3759

c. $P(Z > 1.28) = .1,\ 1.28 = \dfrac{Xi-2.6}{.5},\ Xi = 3.24$

d. $P(Xi > 3) = .2119$ (from part a)

$E[X] = 400(.2119) = 84.76,\ \sigma_X = \sqrt{(400)(.2119)(.7881)} = 8.173$

$P(Z > \dfrac{80-84.76}{8.173}) = P(Z > -.58) = F_Z(.58) = .7190$

e. $P(X \geq 1) = 1 - P(X = 0) = 1 - (.7881)^2 = .3789$

5.86 a. $P(Z < \frac{85-100}{30}) = P(Z < -.5) = .3085$

b. $P(\frac{70-100}{30} < Z < \frac{130-100}{30}) = P(-1 < Z < 1) = 2\,F_z(1) - 1 = .6826$

c. $P(Z > 1.645) = .05$, $1.645 = \frac{Xi-100}{30}$, Xi = 149.35

d. $P(Z > \frac{60-100}{30}) = P(Z > -1.33) = F_Z(1.33) = .9032$

$P(X \geq 1) = 1 - P(X = 0) = 1 - (.0918)^2 = .9916$

e. Use the binomial formula: $P(X = 2) = C_2^4(.9082)^2(.0918)^2 = 0.0417$

f. 90 – 109

g. 130 - 149

5.88 $P(Z > 1.28) = .1$, $1.28 = \frac{130-100}{\sigma}$, $\sigma = 23.4375$

$P(Z > \frac{140-100}{23.4375}) = P(Z > 1.71) = 1 - F_Z(1.71) = .0436$

5.90 $E[X] = 1000(.4) = 400$, $\sigma_x = \sqrt{(1000)(.4)(.6)} = 15.4919$

$P(Z < \frac{500-400}{15.4919}) = P(Z < 6.45) \approx 1.0000$

5.92 The number of calls per 12-hour time period follows a Poisson distribution with $\lambda = 15$ calls/12-hour time period.

Cumulative Distribution Function

```
Poisson with mu = 15.0000
        x       P( X <= x)
     0.00        0.0000
     1.00        0.0000
     2.00        0.0000
     3.00        0.0002
     4.00        0.0009
     5.00        0.0028
     6.00        0.0076
     7.00        0.0180
     8.00        0.0374
     9.00        0.0699
    10.00        0.1185
    11.00        0.1848
    12.00        0.2676
    13.00        0.3632
    14.00        0.4657
    15.00        0.5681
    16.00        0.6641
    17.00        0.7489
```

$P(x < 10) = P(x \leq 9) = 0.0699$

$P(x > 17) = 1 - P(x \leq 17) = 1 - 0.7489 = 0.2511$

5.94 a. $E[X] = 600(.4) = 240$, $\sigma_x = \sqrt{(600)(.4)(.6)} = 12$

$P(Z > \frac{260-240}{12}) = P(Z > 1.67) = 1 - F_Z(1.67) = .0475$

b. $P(Z > -.254) = .6$, $-.254 = \frac{Xi-240}{12}$, $Xi = 236.95$ (237 listeners)

5.96 $P(Z>1.28)=.1$, $1.28=\frac{3.5-2.4}{\sigma}$, $\sigma=.8594$. Probability that 1 exec spends 3+ hours on task: $P(Z > \frac{3-2.4}{.8594}) = P(Z > .7) = 1 - F_Z(.7) = .242$. $E[X] = 400(.242) = 96.8$, $\sigma_x = \sqrt{(400)(.242)(.758)} = 8.566$. $P(Z > \frac{80-96.8}{8.566})=$ $P(Z>-1.96) = F_Z(1.96)=.975$

5.98 Portfolio consists of 10 shares of stock A and 8 shares of stock B

a. Find the mean and variance of the portfolio value: W = 10X + 8Y with correlation of .3.

$\mu_W = a\mu_x + b\mu_y = 10(12) + 8(10) = 200$

$\sigma^2{}_W = a^2\sigma^2{}_X + b^2\sigma^2{}_Y + 2abCorr(X,Y)\sigma_X\sigma_Y$

$= 10^2(14) + 8^2(12) + 2(10)(8)(.5)(3.74166)(3.4641) = 3{,}204.919$

b. Option 1: Stock 1 with mean of 12, variance of 25, correlation of -.2.

$\sigma^2{}_W = a^2\sigma^2{}_X + b^2\sigma^2{}_Y + 2abCorr(X,Y)\sigma_X\sigma_Y$

$= 12^2(25) + 8^2(12) + 2(10)(8)(-.2)(5)(3.4641) = 3{,}813.744$

Option 2: Stock 2 with mean of 10, variance of 9, correlation of .6.

$= 10^2(9) + 8^2(12) + 2(10)(8)(.6)(3)(3.4641) = 2{,}665.66$

To reduce the variance of the porfolio, select Option 2

5.100 a. $\mu_W = a\mu_x + b\mu_y = 1(40) + 1(35) = 75$

$\sigma^2{}_W = a^2\sigma^2{}_X + b^2\sigma^2{}_Y + 2abCorr(X,Y)\sigma_X\sigma_Y$

$= 1^2(100) + 1^2(144) + 2(1)(1)(.6)(10)(12) = 388$

$\sigma_W = \sqrt{388} = 19.69772$

Probability that all seats are filled:

$\frac{100-75}{19.69772} = 1.27$ $Fz = .8980$. $1 - .8980 = .1020$

b. Probability that between 75 and 90 seats will be filled:

$\frac{90-75}{19.69772} = .76$ $.5 - Fz(.76) = .2764$

5.102 Mean and variance for stock prices:

	AB Volvo (ADR)	Alcoa Inc.	Pentair Inc.	TCF Financial Corporation
Mean	8.6143	31.9829	28.9543	25.1643
Variance	25.4171	27.4188	95.4157	20.4361

Covariances:

	AB Volvo (ADR)	Alcoa Inc.	Pentair Inc.
Alcoa Inc.	6.5180		
Pentair Inc.	31.2128	5.4712	
TCF Financial Corporation	-4.3594	-2.7947	20.6897

Let the total value of the portfolio be represented by variable W.

μ_W = (0.3333)(8.6143) + (0.1667)(31.9829) + (0.3333)(28.9543) + (0.1667)(25.1643) = 22.05

σ_W^2 = $(0.3333^2)(25.4171) + (0.1667^2)(27.4188) + (0.3333^2)(95.4157) + (0.1667^2)(20.4361) + 2[(0.3333)(0.1667)(6.5180) + (0.3333)(0.3333)(31.2128) + (0.3333)(0.1667)(-4.3594) + (0.1667)(0.3333)(5.4712) + (0.1667)(0.1667)(-2.7947) + (0.3333)(0.1667)(20.6897)] = 24.68$

We can confirm these results by finding the portfolio price for each year, shown next, and then by finding the mean and variance of these prices.

Portfolio Price
26.1833
24.6217
24.0983
27.7533
20.2700
14.3017
17.1033

Descriptive Statistics: Portfolio Price

```
Variable          N  N*   Mean  SE Mean  StDev  Variance  Minimum     Q1  Median
Portfolio Price   7   0  22.05     1.88   4.97     24.68    14.30  17.10   24.10

Variable            Q3  Maximum
Portfolio Price  26.18    27.75
```

The output above confirms that $\mu_W = 22.05$ and $\sigma_W^2 = 24.68$.

Assuming that the portfolio price is normally distributed, the narrowest interval that contains 95% of the distribution of portfolio value is centered at the mean. Therefore, it is $\mu_W \pm z_{\alpha/2}\sigma_W$. Using $z_{\alpha/2} = 1.96$ and $\sigma_W = 4.97$, the interval is $22.05 \pm (1.96)(4.97)$ or (12.31, 31.79).

5.104 Mean and variance for stock price growth:

	3M Company	Alcoa Inc.	Intel Corporation	Potlatch Corporation	General Motors Corporation	Sea Containers Ltd.
Mean	0.001992	0.004389	-0.000082	0.007449	-0.014355	-0.146323
Variance	0.002704	0.005060	0.006727	0.006674	0.014518	0.176663

Covariances:

	3M Company	Alcoa Inc.	Intel Corporation	Potlatch Corporation	General Motors Corporation
Alcoa Inc.	0.00153782				
Intel Corporation	0.00163165	0.00184360			
Potlatch Corporation	0.00012217	0.00197600	0.00144736		
General Motors Corporation	-0.00005101	0.00103371	-0.00006588	0.00246545	
Sea Containers Ltd.	0.00075015	0.00706908	-0.00131221	-0.00151704	0.01077420

Let the portfolio growth be represented by variable W. The mean and variance for this portfolio, μ_W and σ_W^2, can be found using the following equations or by using technology.

$$\mu_W = \sum_{i=1}^{k} a_i\mu_i\,,\ \sigma_W^2 = \sum_{i=1}^{k} a_i^2\sigma_i^2 + 2\sum_{i=1}^{k-1}\ \sum_{j=i+1}^{k} a_i a_j\, Cov(X_i, X_j)$$

Descriptive Statistics: Portfolio Growth

```
Variable            N  N*     Mean  SE Mean   StDev  Variance  Minimum       Q1
Portfolio Growth   60   0  -0.0245   0.0111  0.0862    0.0074  -0.4182  -0.0688

Variable          Median      Q3  Maximum
Portfolio Growth -0.0062  0.0303   0.1212
```

As previously shown, $\mu_W = -0.0245$ and $\sigma_W^2 = 0.0074$.

5.106 Mean and variance for stock price growth:

	AB Volvo	Pentair Inc.	Reliant Energy Inc.	TCF Financial Corporation	3M Company	Restoration Hardware Inc.
Mean	0.019592	0.007641	0.019031	-0.004087	0.001992	-0.013406
Variance	0.004805	0.006227	0.012686	0.004001	0.002704	0.027618

Covariances:

	AB Volvo	Pentair Inc.	Reliant Energy Inc.	TCF Financial Corporation	3M Company
Pentair Inc.	0.00074848				
Reliant Energy Inc.	0.00228027	0.00105381			
TCF Financial Corporation	-0.00001514	-0.00021080	-0.00041228		
3M Company	0.00099279	0.00087718	0.00031032	0.00072435	
Restoration Hardware Inc.	0.00117969	0.00169410	0.00055922	-0.00041072	0.00204408

Let the portfolio growth be represented by variable W. The mean and variance for this portfolio, μ_W and σ_W^2, can be found using the following equations or by using technology.

$$\mu_W = \sum_{i=1}^{k} a_i \mu_i \, , \; \sigma_W^2 = \sum_{i=1}^{k} a_i^2 \sigma_i^2 + 2\sum_{i=1}^{k-1} \sum_{j=i+1}^{k} a_i a_j \; Cov(X_i, X_j)$$

Descriptive Statistics: Portfolio Growth

```
Variable           N  N*     Mean  SE Mean    StDev  Variance   Minimum
Portfolio Growth  60   0  0.00513  0.00612  0.04740   0.00225  -0.16714

Variable                Q1   Median       Q3  Maximum
Portfolio Growth  -0.02762  0.00631  0.04184  0.10438
```

As previously shown, $\mu_W = 0.00513$ and $\sigma_W^2 = 0.00225$.

For the second portfolio (20% AB Volvo, 30% Pentair, 30% Reliant Energy, and 20% 3M Company), we get the following output:

Descriptive Statistics: Portfolio Growth

```
Variable           N  N*     Mean  SE Mean    StDev  Variance   Minimum
Portfolio Growth  60   0  0.01232  0.00680  0.05270   0.00278  -0.15522

Variable                Q1   Median       Q3  Maximum
Portfolio Growth  -0.02121  0.01386  0.05357  0.10539
```

For the second portfolio, as previously shown, $\mu_W = 0.01232$ and $\sigma_W^2 = 0.00278$. The second portfolio has a higher mean and a higher variance. Recall that risk is directly related to variance. Since the second portfolio has a significantly larger mean and only a slightly larger variance, it would be the better choice.

Chapter 6:

Sampling and Sampling Distributions

6.2 a. Binomial random variable with n = 2, p = .5

Probability Density Function

```
Binomial with n = 2 and p = 0.5
 x   P( X = x )
 0        0.25
 1        0.50
 2        0.25
```

b. Binomial random variable with n = 4, p = .5

Probability Density Function

```
Binomial with n = 4 and p = 0.5
 x   P( X = x )
 0      0.0625
 1      0.2500
 2      0.3750
 3      0.2500
 4      0.0625
```

c. Binomial random variable with n = 10, p = .5

Probability Density Function

```
Binomial with n = 10 and p = 0.5
 x   P( X = x )
 0    0.000977
 1    0.009766
 2    0.043945
 3    0.117188
 4    0.205078
 5    0.246094
 6    0.205078
 7    0.117188
 8    0.043945
 9    0.009766
10    0.000977
```

6.4 The response should note that there will be errors in taking a census of the entire population as well as errors in taking a sample. Improved accuracy can be achieved via sampling methods versus taking a complete census (see reference to Hogan, 90). By using sample information, we can make valid inferences about the entire population without the time and expense involved in taking a census.

6.6 a. mean and variance of the sampling distribution for the sample mean

$\mu_{\bar{x}} = \mu = 100$

$\sigma^2_{\bar{x}} = \sigma^2/n = 900/30 = 30 \quad \sigma_{\bar{x}} = \sqrt{\sigma_{\bar{x}}^2} = \sqrt{30}$

b. Probability that $\bar{x} > 109$ $\quad z_{\bar{x}} = \dfrac{109-100}{\sqrt{30}} = 1.64 \quad 1 - Fz(1.64) = .0505$

c. Probability that $96 \le \bar{x} \le 110$ $z_{\bar{x}} = \dfrac{96-100}{\sqrt{30}} = -.73$ $1 - Fz(.73) = .2327$

$z_{\bar{x}} = \dfrac{110-100}{\sqrt{30}} = 1.83$ Fz = .9664. .9664 - .2327 = .7337

d. Probability that $\bar{x} \le 107$ $z_{\bar{x}} = \dfrac{107-100}{\sqrt{30}} = 1.28$ Fz = .8997

6.8 a. mean and variance of the sampling distribution for the sample mean

$\mu_{\bar{x}} = \mu = 400$

$\sigma^2_{\bar{x}} = \sigma^2 / n = 1600/35 = 45.7143$ $\sigma_{\bar{x}} = \sqrt{\sigma_{\bar{x}}^2} = \sqrt{45.7143}$

b. Probability that $\bar{x} > 412$ $z_{\bar{x}} = \dfrac{412-400}{\sqrt{45.7143}} = 1.77$ $1 - Fz(1.77) = .0384$

c. Probability that $393 \le \bar{x} \le 407$ $z_{\bar{x}} = \dfrac{407-400}{\sqrt{45.7143}} = 1.04$ $Fz(1.04) = .8508$

$z_{\bar{x}} = \dfrac{393-400}{\sqrt{45.7143}} = -1.04$ $1 - Fz(1.04) = .1492.$ $.8508 - .1492 = .7016$

d. Probability that $\bar{x} \le 389$ $z_{\bar{x}} = \dfrac{389-400}{\sqrt{45.7143}} = -1.63$ $1\text{-}Fz(1.63) = 1\text{-}.9484 = .0516$

6.10 a. $E(\bar{X}) = \mu_{\bar{x}} = 1{,}200$

b. $\sigma_{\bar{x}}^2 = \dfrac{\sigma^2}{n} = \dfrac{(400)^2}{9} = 17{,}778 = .1292$

c. $\sigma_{\bar{x}} = \dfrac{\sigma}{\sqrt{n}} = \dfrac{400}{3} = 133.33$

d. $P(Z < \dfrac{1{,}050 - 1{,}200}{133.33}) = P(Z < -1.13)$

6.12 a. $P(\bar{x} > 210{,}000) = P\left(z > \dfrac{210{,}000 - 215{,}000}{25{,}000/\sqrt{100}}\right) = P(z > -2) = 0.9772$

b. $P(213{,}000 < \bar{x} < 217{,}000) = P\left(\dfrac{213{,}000 - 215{,}000}{25{,}000/\sqrt{100}} < z < \dfrac{217{,}000 - 215{,}000}{25{,}000/\sqrt{100}}\right) = P(-0.8 < z < 0.8)$

$= 0.5763$

c. $P(214{,}000 < \bar{x} < 216{,}000) = P\left(\dfrac{214{,}000 - 215{,}000}{25{,}000/\sqrt{100}} < z < \dfrac{216{,}000 - 215{,}000}{25{,}000/\sqrt{100}}\right) = P(-0.4 < z < 0.4)$

$= 0.3108$

d. The sample mean selling price is most likely to lie in the range \$214,000 to \$216,000 since it is centered about the given population mean.

e. The results were still valid because of the central limit theorem with the sample size being larger than 30.

6.14 a. $\sigma_{\bar{x}} = \frac{22}{\sqrt{16}} = 5.5$

b. $P(Z < \frac{100-87}{5.5}) = P(Z < 2.36) = .9909$

c. $P(Z > \frac{80-87}{5.5}) = P(Z > -1.27) = .8980$

d. $P(\frac{85-87}{5.5} > Z > \frac{95-87}{5.5}) = P(-.36 > Z > 1.45) = .4329$

e. Higher, higher, lower. The graph will show that the standard error of the sample means will decrease with an increased sample size.

6.16 a. $\sigma_{\bar{x}} = \frac{40}{\sqrt{100}} = 4$

b. $P(Z > 5/4) = P(Z > 1.25) = .1056$

c. $P(Z < -4/4) = P(Z < -1) = .1587$

d. $P(-3/4 > Z > 3/4) = P(-.75 > Z > .75) = .4532$

6.18 a. $\sigma_{\bar{x}} = \frac{1.6}{\sqrt{100}} = .16$, $P(Z>1.645) = .05$, $1.645 = \frac{Difference}{.16}$, Difference = ±.2632

b. $P(Z < -1.28) = .1$, $-1.28 = \frac{Difference}{.16}$, Difference = –.2048

c. $P(Z > 1.44) = .075$, $1.44 = \frac{Difference}{.16}$, Difference = .2304

6.20 a. $P(Z > 1.96) = .025$, $1.96 = \frac{2}{8.4/\sqrt{n}}$, n = 67.766, take n = 68

b. smaller

c. larger

6.22 a. $N = 20$, correction factor = $\frac{0}{19}$

$N = 40$, correction factor = $\frac{20}{39}$

$N = 100$, correction factor = $\frac{80}{99}$

$N = 1{,}000$, correction factor = $\frac{980}{999}$

$N = 10{,}000$, correction factor = $\frac{9{,}980}{9{,}999}$

b. When the population size (N) equals the sample size (n), then there is no variation away from the population mean and the standard error will be zero. As the sample size becomes relatively small compared to the population size, the correction factor tends towards 1 and the correction factor becomes less significant in the calculation of the standard error

c. The correction factor tends toward a value of 1 and becomes progressively less important as a modifying factor when the sample size decreases relative to the population size

6.24 $\sigma_{\bar{x}} = \frac{30}{\sqrt{50}}\sqrt{\frac{200}{249}} = 3.8023$

a. $P(Z > \frac{2.5}{3.8023}) = P(Z > .66) = .2546$

b. $P(Z < \frac{-5}{3.8023}) = P(Z < -1.31) = .0951$

c. $P(\frac{-10}{3.8023} < Z < \frac{10}{3.8023}) = P(-2.63 < Z < 2.63) = 1 - .9914 = .0086$

6.26 $E(\hat{p}) = .4 \quad \sigma_{\hat{p}} = \sqrt{\frac{(.4)(.6)}{100}} = .04899$

a. Probability that the sample proportion is greater than .45

$z = \frac{.45 - .4}{.04899} = P(Z > 1.02) = .1539$

b. Probability that the sample proportion is less than .29

$z = \frac{.29 - .4}{.04899} = P(Z < -2.25) = .0122$

c. Probability that the sample proportion is between .35 and .51

d. $P(\frac{.35 - .4}{.04899} < Z < \frac{.51 - .4}{.04899}) = P(-1.02 < Z < 2.25) = .8339$

6.28 $E(\hat{p}) = .60 \quad \sigma_{\hat{p}} = \sqrt{\frac{(.6)(.4)}{100}} = .04899$

a. Probability that the sample proportion is greater than .66

$z = \frac{.66 - .6}{.04899} = P(Z > 1.22) = .1112$

b. Probability that the sample proportion is less than .48

$z = \frac{.48 - .6}{.04899} = P(Z < -2.45) = .0071$

c. Probability that the sample proportion is between .52 and .66

$P(z = \frac{.52 - .6}{.04899} < Z < z = \frac{.66 - .6}{.04899}) = P(-1.63 < Z < 1.22) = .8372$

6.30 a. $E(\hat{p}) = .424$

b. $\sigma_{\hat{p}}^2 = \frac{(.424)(.576)}{100} = .00244$

c. $\sigma_{\hat{p}} = .0494$

d. $P(Z > \frac{.5 - .424}{.0494}) = P(Z > 1.54) = .0618$

6.32 a. $E(\hat{p}) = .20$

b. $\sigma_{\hat{p}}^2 = \frac{(.2)(.8)}{180} = .000889$

c. $\sigma_{\hat{p}} = .0298$

d. $P(Z < \frac{.15 - .2}{.0298}) = P(Z < -1.68) = .0465$

6.34 $\sigma_{\hat{p}} = \sqrt{\frac{(.4)(.6)}{120}} = .0447$

$P(\frac{.35 - .4}{.0447} < Z < \frac{.45 - .4}{.0447}) = P(-1.12 < Z < 1.12) = .7372$

6.36 a. $\sigma_{\hat{p}} = \sqrt{\frac{(.2)(.8)}{130}} = .0351$

b. $(Z > \frac{.15 - .2}{.0351}) = P(Z > -1.42) = .9222$

c. $P(\frac{.18 - .2}{.0351} < Z < \frac{.22 - .2}{.0351}) = P(-.57 < Z < .57) = .4314$

d. Higher, higher

6.38 The largest value for $\sigma_{\hat{p}}$ is when p = .5. In this case, $\sigma_{\hat{p}} = \sqrt{\frac{(.5)(.5)}{100}} = .05$

6.40 a. $\sigma_{\hat{p}} = \sqrt{\frac{(.25)(.75)}{120}} = .0395$

b. $P(Z > 1.28)$, $1.28 = \frac{Difference}{.0395}$, Difference = .0506

c. $P(Z < -1.645)$, $-1.645 = \frac{Difference}{.0395}$, Difference = .065

d. $P(Z > 1.036)$, $1.036 = \frac{Difference}{.0395}$, Difference = .0409

6.42 $\sigma_{\hat{p}} = \sqrt{\frac{(.5)(.5)}{250}} = .03162$, P(Z > $\frac{.58-.5}{.03162}$)= P(Z > 2.53) = .0057

6.44 a. $\sigma_{\hat{p}} = \sqrt{\frac{0.4(1-0.4)}{120}} = 0.0447$

b. $P(\hat{p} < 0.33) = P\left(z < \frac{0.33-0.40}{0.0447}\right) = P(z < -1.57) = 0.0582$

c. $P(0.38 < \hat{p} < 0.46) = P\left(\frac{0.38-0.40}{0.044721} < z < \frac{0.46-0.40}{0.044721}\right)$
$= P(-0.4472 < z < 1.34) = 0.5825$

6.46 P(Z < $\frac{.1-.122}{.036}$ = P(Z < -.61) = .2709, $\sigma_{\hat{p}} = \sqrt{\frac{(.2709)(.7291)}{81}} = .04969$

P(Z > $\frac{.5-.2709}{.04969}$) = P(Z > 4.61) ≈ .0000

6.48 a. Probability that the sample mean is > 200.

Probability that $\bar{x} > 200$ $z_{\bar{x}} = \frac{200-198}{10/\sqrt{25}} = 1.00$ 1 – Fz(1.00) = .1587

b. 5% of the sample variances would be less than this value

$P(s^2 > k) = P\left[\frac{(n-1)s^2}{\sigma^2}\right]$ $\chi^2_{24,.95} = 13.85$ $\frac{24s^2}{100} < 13.85$ $s^2 < 57.702$

c.5% of the samples variances would be greater than this value

$P(s^2 > k) = P\left[\frac{(n-1)s^2}{\sigma^2}\right]$ $\chi^2_{24,.05} = 36.42$ $\frac{24s^2}{100} > 36.42$ $s^2 > 151.879$

6.50 $P(\frac{(n-1)s^2}{\sigma^2} > \frac{19(3.1)}{1.75}) = P(\chi^2_{(19)} > 33.66)$ = between .01 and .025 (.0201 exactly)

6.52 a. $P(\frac{(n-1)s^2}{\sigma^2} > \frac{15(3,000)^2}{(2,500)^2}) = P(\chi^2_{(15)} > 21.6)$ = greater than .1 (.1187 exactly)

b. $P(\frac{(n-1)s^2}{\sigma^2} < \frac{15(1,500)^2}{(2,500)^2}) = P(\chi^2_{(15)} < 5.4)$ = between .01 and .025 (.0118 exactly)

6.54 a. $P(\frac{(n-1)s^2}{\sigma^2} < \frac{24(75)^2}{(100)^2}) = P(\chi^2_{(24)} < 13.5)$ = between .025 and .05 (.0428 exactly)

b. $P(\frac{(n-1)s^2}{\sigma^2} > \frac{24(150)^2}{(100)^2}) = P(\chi^2_{(24)} > 54)$ = less than .005 (.0004 exactly)

6.56 **Descriptive Statistics: C20, C21, C22, C23, C24, C25, C26, C27, ...**

Variable	Mean	Variance
C20	3.00	2.00
C21	4.00	8.00
C22	4.00	8.00
C23	4.50	12.50
C24	5.00	18.00
C25	5.00	2.00
C26	5.00	2.00
C27	6.00	8.00
C28	6.0000	0.000000000
C29	6.500	0.500
C30	7.00	2.00
C31	6.500	0.500
C32	7.00	2.00
C33	7.500	0.500
C34	5.50	4.50

Descriptive Statistics: Variance

Variable	Mean	StDev	Variance	Sum
Variance	4.72	5.26	27.62	70.80

$\overline{x} = \dfrac{70.8}{15} = 4.72 \quad E(s^2) = {15(3.91667)}\Big/{(14)} = 4.1964$

6.58 a. $P(\chi^2_{(9)} > 14.68) = .10$, $14.68 = 9(\text{Difference})$, Difference = 1.6311 (163.11%)

b. $P(\chi^2_{(9)} < 2.7) = .025$, $P(\chi^2_{(9)} > 19.02) = .025$,
$2.7 = 9\text{a}$, a = .3, $19.02 = 9\text{b}$, b = 2.1133
The probability is .95 that the sample variance is between 30% and 211.33% of the population variance

c. The interval in part b. will be smaller

6.60 a. $P(\chi^2_{(11)} > 4.57) = .95$, $4.57 = 11\text{Difference}$, Difference = .4155 (41.55%)

b. $P(\chi^2_{(11)} > 5.58) = .90$, $5.58 = 11\text{Difference}$, Difference = .5073 (50.73%)

c. $P(\chi^2_{(11)} < 3.82) = .025$, $P(\chi^2_{(11)} > 21.92) = .025$,
$3.82 = 11\text{a}$, a = .34727, $21.92 = 11\text{b}$, b = 1.9927
The probability is .95 that the sample variance is between 34.727% and 199.27% of the population variance

6.62 $P\left(\dfrac{(n-1)s^2}{\sigma^2} < \dfrac{24(12.2)}{15.4}\right) = P(\chi^2_{(24)} < 19.01)$ = less than .90 (.5438 exactly)

6.64 a. $C_2^6 = \dfrac{6!}{2!4!} = 15$ possible samples

b. (41, 39), (41, 35), (41, 35), (41, 33), (41, 38), (39, 35), (39, 35), (39, 33), (39, 38), (35, 35), (35, 33), (35, 38), (35, 33), (35, 38), (33, 38)

c. $34P_{\overline{X}}(34) = 34\frac{2}{15} = 4.5333$

$35P_{\overline{X}}(35) = \frac{35}{15} = 2.3333$

$35.5P_{\overline{X}}(35.5) = \frac{35.5}{15} = 2.3667$

$36P_{\overline{X}}(36) = \frac{36}{15} = 2.4$

$36.5P_{\overline{X}}(36.5) = 36.5\frac{2}{15} = 4.8667$

$37P_{\overline{X}}(37) = 37\frac{3}{15} = 7.4$

$38P_{\overline{X}}(38) = 38\frac{2}{15} = 5.0667$

$38.5P_{\overline{X}}(38.5) = \frac{38.5}{15} = 2.5667$

$39.5P_{\overline{X}}(39.5) = \frac{39.5}{15} = 2.6333$

$40P_{\overline{X}}(40) = \frac{40}{15} = 2.6667$

d. The mean of the sampling distribution of the sample mean is $\sum \overline{x}P_{\overline{x}}(\overline{x}) = 36.8333$ which is exactly equal to the population mean: $\frac{1}{N}\sum x_i = 36.8333$. This is the result expected from the Central Limit Theorem.

6.66 a. $P(Z > \frac{450-420}{100/\sqrt{25}}) = P(Z > 1.5) = .0668$

b. $P(\frac{400-420}{100/\sqrt{25}} < Z < \frac{450-420}{100/\sqrt{25}}) = P(-1 < Z < 1.5) = .7745$

c. $P(Z > 1.28) = .1,\ 1.28 = \frac{\overline{x}-420}{100/\sqrt{25}},\ \overline{x} = 445.6$

d. $P(Z < -1.28) = .1,\ -1.28 = \frac{\overline{x}-420}{100/\sqrt{25}},\ \overline{x} = 394.4$

e. $P(\chi^2_{(24)} > 36.42) = .05,\ 36.42 = \frac{24s^2}{(100)^2},\ s = 123.1868$

f. $P(\chi^2_{(24)} < 13.85) = .05,\ 13.85 = \frac{24s^2}{(100)^2},\ s = 75.966$

g. Smaller. A larger sample size would lead to a smaller standard error and the graph of the normal distribution would be tighter with less area in the tails.

6.68 a. $P(Z > \frac{19-14.8}{6.3/\sqrt{9}}) = P(Z > 2) = .0228$

b. $P(\frac{10.6-14.8}{6.3/\sqrt{9}} < Z < \frac{19-14.8}{6.3/\sqrt{9}}) = P(-2 < Z < 2) = .9544$

c. $P(Z < -.675) = .25$, $-.675 = \frac{X_i - 14.8}{6.3/\sqrt{9}}$, $X_i = 13.3825$

d. $P(\chi^2_{(8)} > 13.36) = .1$, $13.36 = \frac{8s^2}{(6.3)^2}$, $s = 8.1414$

e. Smaller

6.70 Let n = N, then $\bar{X} = \mu_x$:

$$E[\sum_{i=1}^{N}(X_i - \bar{X})^2] = n\sigma^2_x - n\frac{\sigma^2_x}{n}\frac{N-n}{N-1} = n\sigma^2_x - \frac{N-n}{N-1}\sigma^2_x =$$

$$\frac{\sigma^2_x}{N-1}(nN - n - N + n) = \frac{N\sigma^2_x}{N-1}(n-1)$$

Therefore, $E[\frac{1}{n-1}\sum(X_i - \bar{X})^2] = \frac{1}{n-1}E[\sum(X_i - \bar{X})^2] = \frac{N\sigma^2_x}{N-1}$

6.72 a. $P(\hat{p} < 0.7) = P\left(z < \frac{0.7-0.8}{\sqrt{(0.8)(0.2)/60}}\right) = P(z < -1.94) = 0.0262$

b. $P(\hat{p} < 0.7) = P\left(z < \frac{0.7-0.8}{\sqrt{(0.8)(0.2)/6}}\right) = P(z < -0.61) = 0.2709$

c. $P(\bar{x} > 38{,}000) = P\left(z > \frac{38{,}000-37{,}000}{4{,}000/\sqrt{6}}\right) = P(z > 0.61) = 0.2709$

d. $P(x > 38{,}000) = P\left(z > \frac{38{,}000-37{,}000}{4{,}000}\right) = P(z > 0.25) = 0.5987$

6.74 $P(\frac{(n-1)s^2}{\sigma^2} > 20(2)) = P(\chi^2_{(20)} > 40) = .005$

6.76 $10 < \mu_{\bar{x}} < \bar{X} + 10$, $-10 < \bar{X} - \mu_{\bar{x}} < 10$

$P(\frac{-10}{40/\sqrt{16}} < Z < \frac{10}{40/\sqrt{16}}) = P(-1 < Z < 1) = .6826$

6.78 a. $\sigma_{\bar{x}} = \sqrt{\frac{(.4)(.6)}{250}} = .03098$

$P(Z > -.843) = .8,\ -.843 = \frac{p - .4}{.03098},\ p = .3739$

b. $P(Z < 1.28) = .9,\ 1.28 = \frac{p - .4}{.03098},\ p = .4397$

c. $P(Z > 1.04) = .35,\ 1.04 = \frac{Difference}{.03098}$, Difference = ±.0322

6.80 a. $P(\frac{(n-1)s^2}{\sigma^2} > \frac{24(4,000)^2}{(6,600)^2}) = P(\chi^2_{(24)} > 8.82)$ = more than .99 (.9979 exactly)

b. $P(\frac{(n-1)s^2}{\sigma^2} < \frac{24(8,000)^2}{(6,600)^2}) = P(\chi^2_{(24)} < 35.62)$ = between .9 and .95 (.9354 exactly)

6.82 a. The sample mean is $x = \frac{\sum x}{n} = \frac{70803}{100} = 708.03$.

The sample standard deviation is $s = 8.106$.

To find the standard deviation of the sample mean for each sample, use the equation $\sigma_{\bar{x}} = \frac{\sigma}{\sqrt{n}}$. Use $s = 8.106$ as an estimate for σ and note that $n = 5$ for each sample.

Thus, $\sigma_{\bar{x}} = \frac{8.106}{\sqrt{5}} = 3.625$.

b. $P(\bar{x} < 685) = P\left(\frac{\bar{x} - \mu}{\sigma_{\bar{x}}} < \frac{685 - 710}{3.625}\right) = P(z < -6.90) = 0.0000$

c. $P(\bar{x} > 720) = P\left(\frac{\bar{x} - \mu}{\sigma_{\bar{x}}} > \frac{720 - 710}{3.625}\right) = P(z > 2.76) = 0.0029$

6.84 a. The sample mean is $x = \frac{\sum x}{n} = \frac{156654.6}{138} = 1135.178$.

The sample variance is $s^2 = 11.3382$.

To find the variance of the sample mean for each sample, use the equation $\sigma_{\bar{x}} = \frac{\sigma}{\sqrt{n}}$, which gives us $\sigma_{\bar{x}}^2 = \frac{\sigma^2}{n}$.

Use $s^2 = 11.3382$ as an estimate for σ^2 and note that $n = 6$ for each sample. Thus, $\sigma_{\bar{x}}^2 = \frac{11.3382}{6} = 1.8898$.

b. Use $\sigma_{\bar{x}} = \sqrt{1.8898} = 1.3747$.

$$P(1120 < \bar{x} < 1150) = P\left(\frac{1120-1134}{1.8898} < \frac{\bar{x}-\mu}{\sigma_{\bar{x}}} < \frac{1150-1134}{1.8898}\right)$$

$$= P(-10.18 < z < 11.64) = 1.0000$$

Chapter 7:

Estimation: Single Population

7.2 a. There appears to be no evidence of non-normality, as shown by the normal probability plot shown next.

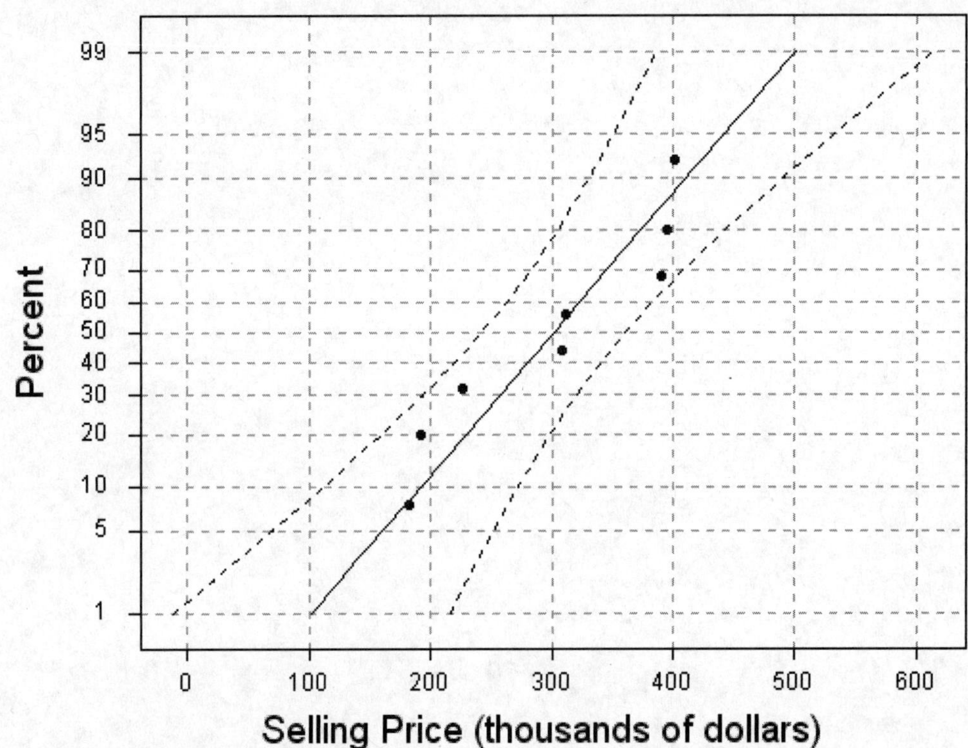

b. Assuming normality, the unbiased and most efficient point estimator for the population mean is the sample mean, $\overline{X}$.

$$\overline{X} = \frac{\sum X_i}{n} = \frac{2411}{8} = 301.375 \text{ thousand dollars}$$

c. Assuming normality, the sample mean is an unbiased estimator of the population mean with variance $\text{Var}(\overline{X}) = \frac{\sigma^2}{n}$. Also, assuming normality, the unbiased and most efficient point estimator for the population variance is the sample variance, s^2.

$$\text{Var}(\overline{X}) = \frac{s^2}{n} = \frac{8373.125}{8} = 1046.64$$

d. The unbiased and most efficient point estimator for a proportion is the sample proportion, $\hat{p}$.

$$\hat{p} = \frac{x}{n} = \frac{3}{8} = 0.375$$

7.4 n = 12 employees. Number of hours of overtime worked in the last month:

a. Unbiased point estimator of the population mean is the sample mean:

$$\overline{X} = \frac{\sum X_i}{n} = 24.42$$

b. The unbiased point estimate of the population variance: $s^2 = 85.72$

c. Unbiased point estimate of the variance of the sample mean

$$Var(\overline{X}) = \frac{s^2}{n} = \frac{85.72}{12} = 7.1433$$

d. Unbiased estimate of the population proportion: $\hat{p} = \frac{x}{n} = \frac{3}{12} = .25$

e. Unbiased estimate of the variance of the sample proportion:

$$Var(\hat{p}) = \frac{\hat{p}(1-\hat{p})}{n} = \frac{.25(1-.25)}{12} = .015625$$

7.6 a. $E(\overline{X}) = \frac{1}{2}E(X_1) + \frac{1}{2}E(X_2) = \frac{\mu}{2} + \frac{\mu}{2} = \mu$

$$E(Y) = \frac{1}{4}E(X_1) + \frac{3}{4}E(X_2) = \frac{\mu}{4} + \frac{3\mu}{4} = \mu$$

$$E(Z) = \frac{1}{3}E(X_1) + \frac{2}{3}E(X_2) = \frac{\mu}{3} + \frac{2\mu}{3} = \mu$$

b. $Var(\overline{X}) = \frac{\sigma^2}{n} = \frac{1}{4}Var(X_1) + \frac{1}{4}Var(X_2) = \frac{1}{2}\frac{\sigma^2}{8} = \frac{\sigma^2}{4}$

$$Var(Y) = \frac{1}{16}Var(X_1) + \frac{9}{16}Var(X_2) = \frac{5\sigma^2}{8}$$

$$Var(Z) = \frac{1}{9}Var(X_1) + \frac{4}{9}Var(X_2) = \frac{5\sigma^2}{9}$$

$\overline{X}$ is most efficient since $Var(\overline{X}) < Var(Y) < Var(Z)$

c. Relative efficiency between Y and $\overline{X}$: $\frac{Var(Y)}{Var(\overline{X})} = \frac{5}{2} = 2.5$

Relative efficiency between Z and $\overline{X}$: $\frac{Var(Z)}{Var(\overline{X})} = \frac{20}{9} = 2.222$

7.8 a. Evidence of non-normality?
No evidence of the data distribution coming from a non-normal population
b. The minimum variance unbiased point estimator of the population mean is the sample mean: $\bar{X} = \frac{\sum X_i}{n} = 3.8079$

Descriptive Statistics: Volumes

Variable	N	Mean	Median	TrMean	StDev	SE Mean
Volumes	75	3.8079	3.7900	3.8054	0.1024	0.0118

Variable	Minimum	Maximum	Q1	Q3
Volumes	3.5700	4.1100	3.7400	3.8700

c. Minimum variance unbiased point estimate of the population variance is the sample variance $s^2 = .0105$

7.10 Calculate the margin of error to estimate the population mean
a. 98% confidence level; n = 64, variance = 144

$$ME = z_{\alpha/2} \sigma/\sqrt{n} = 2.33\left(12/\sqrt{64}\right) = 3.495$$

b. 99% confidence interval, n=120; standard deviation = 100

$$ME = z_{\alpha/2} \sigma/\sqrt{n} = 2.58\left(100/\sqrt{120}\right) = 23.552$$

7.12 Calculate the LCL and UCL

a. $\bar{x} \pm z_{\alpha/2} \sigma/\sqrt{n} = 50 \pm 1.96\left(40/\sqrt{64}\right) = 40.2 \text{ to } 59.8$

b. $\bar{x} \pm z_{\alpha/2} \sigma/\sqrt{n} = 85 \pm 2.58\left(20/\sqrt{225}\right) = 81.56 \text{ to } 88.44$

c. $\bar{x} \pm z_{\alpha/2} \sigma/\sqrt{n} = 510 \pm 1.645\left(50/\sqrt{485}\right) = 506.2652 \text{ to } 513.73478$

7.14 a. Find the reliability factor for 92% confidence level: $z_{\alpha/2} = +/- 1.75$
b. Calculate the standard error of the mean

$$\sigma/\sqrt{n} = 6/\sqrt{90} = .63246$$

c. Calculate the width of the 92% confidence interval for the population mean

$$\text{width} = 2\text{ME} = 2\left[z_{\alpha/2} \sigma/\sqrt{n}\right] = 2\left[1.75\left(6/\sqrt{90}\right)\right] = 2.2136$$

7.16 a. $n = 16, \quad \bar{x} = 4.07, \quad \sigma = .12, \quad z_{.005} = 2.58$

$4.07 \pm 2.58(.12/4) = 3.9926$ up to 4.1474

b. narrower since the z score for a 95% confidence interval is smaller than the z score for the 99% confidence interval

c. narrower due to the smaller standard error

d. wider due to the larger standard error

7.18 Find the ME

a. n = 20, 90% confidence level, s = 36

$$ME = t_{\alpha/2} \, s/\sqrt{n} = ME = 1.729\left(36/\sqrt{120}\right) = 13.9182$$

b. n = 7, 98% confidence level, s = 16

$$ME = t_{\alpha/2} \, s/\sqrt{n} = ME = 3.143\left(16/\sqrt{7}\right) = 19.007$$

c. n = 16, 99% confidence level, $x_1 = 15$, $x_2 = 17$, $x_3 = 13$, $x_4 = 11$, $x_5 = 14$

$$\bar{x} = 14, \quad s = 2.58199 \quad ME = 5.841\left(2.58199/\sqrt{4}\right) = 7.5407$$

7.20 Find the LCL and UCL for each of the following:

a. alpha = .05, n = 25, sample mean = 560, s = 45

$$\bar{x} \pm t_{\alpha/2}\left(s/\sqrt{n}\right) = 560 \pm 2.064\left(45/\sqrt{25}\right) = 541.424 \text{ to } 578.576$$

b. alpha/2 = .05, n = 9, sample mean = 160, sample variance = 36

$$\bar{x} \pm t_{\alpha/2}\left(s/\sqrt{n}\right) = 160 \pm 1.860\left(6/\sqrt{9}\right) = 156.28 \text{ to } 163.72$$

c. $1-\alpha$ = .98, n = 22, sample mean = 58, s = 15

$$\bar{x} \pm t_{\alpha/2}\left(s/\sqrt{n}\right) = 58 \pm 2.518\left(15/\sqrt{22}\right) = 49.9474 \text{ to } 66.0526$$

7.22 Calculate the width for each of the following:

a. alpha = 0.05, n = 6, s = 40

$$w = 2ME = 2t_{\alpha/2} \, s/\sqrt{n} = 2 \cdot 2.571\left(40/\sqrt{6}\right) = 2(41.98425) = 83.9685$$

b. alpha = 0.01, n = 22, sample variance = 400

$$w = 2ME = 2t_{\alpha/2} \, s/\sqrt{n} = 2 \cdot 2.831\left(20/\sqrt{22}\right) = 2(12.07142) = 24.1428$$

c. alpha = 0.10, n = 25, s = 50

$$w = 2ME = 2t_{\alpha/2} \, s/\sqrt{n} = 2 \cdot 1.711\left(50/\sqrt{25}\right) = 2(17.11) = 34.22$$

7.24 a.

Results for: Sugar.xls
Descriptive Statistics: Weights

```
Variable              N         Mean       Median       TrMean         StDev     SE Mean
Weights             100       520.95       518.75       520.52          9.45        0.95

Variable        Minimum      Maximum           Q1           Q3
Weights          504.70       544.80       513.80       527.28
```

90% confidence interval:

Results for: Sugar.xls
One-Sample T: Weights

```
Variable          N        Mean      StDev   SE Mean          90.0% CI
Weights         100     520.948      9.451     0.945   ( 519.379, 522.517)
```

b. narrower since a smaller value of z will be used in generating the 80% confidence interval.

7.26 $n = 7,\quad \overline{x} = 74.7143,\quad s = 6.3957,\quad t_{6,.025} = 2.447$

margin of error: $\pm\ 2.447(6.3957/\sqrt{7}) = \pm\ 5.9152$

7.28 $n = 25,\quad \overline{x} = 42,740,\quad s = 4,780,\quad t_{24,.05} = 1.711$

42,740 ± 1.711(4780/5) = \$41,104.28 up to \$44,375.72

7.30 Find the standard error of the proportion for

a. n = 250, $\hat{p} = 0.3$ $\sqrt{\frac{\hat{p}(1-\hat{p})}{n}} = \sqrt{\frac{.3(.7)}{250}} = .02898$

b. n = 175, $\hat{p} = 0.45$ $\sqrt{\frac{\hat{p}(1-\hat{p})}{n}} = \sqrt{\frac{.45(.55)}{175}} = .03761$

c. n = 400, $\hat{p} = 0.05$ $\sqrt{\frac{\hat{p}(1-\hat{p})}{n}} = \sqrt{\frac{.05(.95)}{400}} = .010897$

7.32 Find the confidence level for estimating the population proportion for

a. 92.5% confidence level; n = 650, $\hat{p} = .10$

$$\hat{p} \pm z_{\alpha/2}\sqrt{\frac{\hat{p}(1-\hat{p})}{n}} = .10 \pm 1.78\sqrt{\frac{.10(1-.10)}{650}} = .079055 \text{ to } .120945$$

b. 99% confidence level; n = 140, $\hat{p} = .01$

$$\hat{p} \pm z_{\alpha/2}\sqrt{\frac{\hat{p}(1-\hat{p})}{n}} = .01 \pm 2.58\sqrt{\frac{.01(1-.01)}{140}} = 0.0 \text{ to } .031696$$

c. alpha = .09; n = 365, $\hat{p} = .50$

$$\hat{p} \pm z_{\alpha/2}\sqrt{\frac{\hat{p}(1-\hat{p})}{n}} = .50 \pm 1.70\sqrt{\frac{.5(.5)}{650}} = .4555 \text{ to } .5445$$

7.34 $n = 95, \quad \hat{p} = 67/95 = .7053, \quad z_{.005} = 2.58$

$$\hat{p} \pm z_{\alpha/2}\sqrt{\frac{\hat{p}(1-\hat{p})}{n}} = .7053 \pm (2.58)\sqrt{\frac{.7053(.2947)}{95}} =$$

99% confidence interval: .5846 up to .8260

7.36 $n = 320, \quad \hat{p} = 80/320 = .25, \quad z_{.025} = 1.96$

$$\hat{p} \pm z_{\alpha/2}\sqrt{\frac{\hat{p}(1-\hat{p})}{n}} = .25 \pm (1.96)\sqrt{.25(.75)/320} =$$

95% confidence interval: .2026 up to .2974

7.38 $width = .545\text{-}.445 = .100; \; ME = 0.05 \; \hat{p} = 0.495 \; \sqrt{\frac{\hat{p}(1-\hat{p})}{n}} = \sqrt{\frac{.495(.505)}{198}} = .0355$

$.05 = z_{\alpha/2}\,(.0355), \quad z_{\alpha/2} = 1.41$

$\alpha = 2[1 - F_z(1.41)] = .0793$

$100(1\text{-}.1586)\% = 84.14\%$

7.40 $n = 246, \quad \hat{p} = 40/246 = .1626, \quad z_{.01} = 2.326$

$$\hat{p} \pm z_{\alpha/2}\sqrt{\frac{\hat{p}(1-\hat{p})}{n}} = .1626 \pm (2.326)\sqrt{.1626(.8374)/246} =$$

98% confidence interval: .1079 up to .2173

7.42 a. $n = 21, \quad s^2 = 16, \quad taking \quad \alpha = .05, \quad \chi^2_{20,.975} = 9.59, \quad \chi^2_{20,.025} = 34.17$

$$\frac{(n-1)s^2}{\chi^2_{n-1,\alpha/2}} < \sigma^2 < \frac{(n-1)s^2}{\chi^2_{n-1,1-\alpha/2}} = \frac{21(16)}{34.17} < \sigma^2 < \frac{21(16)}{9.59} = 9.8332 < \sigma^2 < 35.036$$

b. $n = 16, \quad s = 8, \quad \chi^2_{15,.975} = 6.26, \quad \chi^2_{15,.025} = 27.49$

$$\frac{(n-1)s^2}{\chi^2_{n-1,\alpha/2}} < \sigma^2 < \frac{(n-1)s^2}{\chi^2_{n-1,1-\alpha/2}} = \frac{15(8)^2}{27.49} < \sigma^2 < \frac{15(8)^2}{6.26} = 34.9218 < \sigma^2 < 153.3546$$

c. $n = 28, \quad s = 15, \quad \chi^2_{27,.995} = 11.81, \quad \chi^2_{27,.005} = 49.64$

$$\frac{(n-1)s^2}{\chi^2_{n-1,\alpha/2}} < \sigma^2 < \frac{(n-1)s^2}{\chi^2_{n-1,1-\alpha/2}} = \frac{28(15)^2}{49.64} < \sigma^2 < \frac{28(15)^2}{11.81} = 126.9138 < \sigma^2 < 533.446$$

7.44 Random sample from a normal population

a. Find the 90% confidence interval for the population variance

Descriptive Statistics: Ex7.44

Variable	Mean	SE Mean	StDev	Variance	Minimum	Maximum
Ex7.44	11.00	1.41	3.16	10.00	8.00	16.00

$n = 5,\quad s^2 = 10,\quad \chi^2_{4,.95} = .711,\quad \chi^2_{4,.05} = 9.49$

$$\frac{(n-1)s^2}{\chi^2_{n-1,\alpha/2}} < \sigma^2 < \frac{(n-1)s^2}{\chi^2_{n-1,1-\alpha/2}} = \frac{4(10)}{9.49} < \sigma^2 < \frac{4(10)}{.711} = 4.21496 < \sigma^2 < 56.25879$$

b. Find the 95% confidence interval for the population variance

$n = 5,\quad s^2 = 10,\quad \chi^2_{4,.975} = .484,\quad \chi^2_{4,.025} = 11.14$

$$\frac{(n-1)s^2}{\chi^2_{n-1,\alpha/2}} < \sigma^2 < \frac{(n-1)s^2}{\chi^2_{n-1,1-\alpha/2}} = \frac{4(10)}{11.14} < \sigma^2 < \frac{4(10)}{.484} = 3.59066 < \sigma^2 < 82.6446$$

7.46 $n = 10,\quad s^2 = 28.898,\quad \chi^2_{9,.05} = 16.92,\quad \chi^2_{9,.95} = 3.33$

$$\frac{(n-1)s^2}{\chi^2_{n-1,\alpha/2}} < \sigma^2 < \frac{(n-1)s^2}{\chi^2_{n-1,1-\alpha/2}} =$$

$$\frac{9(28.898)}{16.92} < \sigma^2 < \frac{9(28.898)}{3.33} = 15.3713 \text{ up to } 78.1027$$

7.48 $n = 18,\quad s^2 = 108.16,\quad \chi^2_{17,.05} = 27.59,\quad \chi^2_{17,.95} = 8.67$

$$\frac{(n-1)s^2}{\chi^2_{n-1,\alpha/2}} < \sigma^2 < \frac{(n-1)s^2}{\chi^2_{n-1,1-\alpha/2}} = \frac{17(108.16)}{27.59} < \sigma^2 < \frac{17(108.16)}{8.67} = 66.6444 \text{ up to } 212.0784.$$

Assume that the population is normally distributed.

7.50 $n = 9,\quad s^2 = .7875,\quad \chi^2_{8,.05} = 15.51,\quad \chi^2_{8,.95} = 2.73$

$$\frac{8(.7875)}{15.51} < \sigma^2 < \frac{8(.7875)}{2.73} = .4062 \text{ up to } 2.3077$$

7.52 a. The confidence interval is $\bar{x} - t_{n-1,\alpha/2}\hat{\sigma}_{\bar{x}} < \mu < \bar{x} + t_{n-1,\alpha/2}\hat{\sigma}_{\bar{x}}$. Use $t_{79,0.025} = 1.990$, and $\hat{\sigma}^2_{\bar{x}} = 1.1676$ from exercise 7.51 part a.

$\bar{x} - t_{n-1,\alpha/2}\hat{\sigma}_{\bar{x}} < \mu < \bar{x} + t_{n-1,\alpha/2}\hat{\sigma}_{\bar{x}}$

$142 - (1.990)(1.1676) < \mu < 142 + (1.990)(1.1676)$

or (139.68, 144.32)

b. The confidence interval is $\bar{x} - t_{n-1,\alpha/2}\hat{\sigma}_{\bar{x}} < \mu < \bar{x} + t_{n-1,\alpha/2}\hat{\sigma}_{\bar{x}}$. Use $t_{89,0.025} = 1.987$, and $\hat{\sigma}^2_{\bar{x}} = 0.6667$ from exercise 7.51 part b.

$\bar{x} - t_{n-1,\alpha/2}\hat{\sigma}_{\bar{x}} < \mu < \bar{x} + t_{n-1,\alpha/2}\hat{\sigma}_{\bar{x}}$

$232.4 - (1.987)(0.6667) < \mu < 232.4 + (1.987)(0.6667)$

or (231.08, 233.72)

c. The confidence interval is $\bar{x}-t_{n-1,\alpha/2}\hat{\sigma}_{\bar{x}}<\mu<\bar{x}+t_{n-1,\alpha/2}\hat{\sigma}_{\bar{x}}$. Use $t_{199,0.025}=1.972$, and $\hat{\sigma}_{\bar{x}}^2=0.6049$ from exercise 7.51 part c.

$\bar{x}-t_{n-1,\alpha/2}\hat{\sigma}_{\bar{x}}<\mu<\bar{x}+t_{n-1,\alpha/2}\hat{\sigma}_{\bar{x}}$

$59.3-(1.972)(0.6049)<\mu<59.3+(1.972)(0.6049)$

or (58.11, 60.49)

7.54 a. A $100(1-\alpha)\%$ confidence interval for the population total is obtained from the following formula.

$\hat{p}-z_{\alpha/2}\hat{\sigma}_{\hat{p}}<P<\hat{p}+z_{\alpha/2}\hat{\sigma}_{\hat{p}}$

As stated, $N=1058$, $n=160$, and $x=40$. For a 95% confidence level, note that $z_{\alpha/2}=z_{0.025}=1.96$. Next calculate $\hat{p}$ and $\hat{\sigma}_{\hat{p}}$.

$$\hat{p}=\frac{x}{n}=\frac{40}{160}=0.25$$

$$\hat{\sigma}_{\hat{p}}^2=\frac{\hat{p}(1-\hat{p})}{n-1}\times\frac{(N-n)}{(N-1)}=\frac{0.25(1-0.25)}{160-1}\times\frac{1058-160}{1058-1}=0.0009956$$

$\hat{\sigma}_{\hat{p}}=\sqrt{0.0009956}=0.03155$

So the 95% confidence interval is

$\hat{p}-z_{\alpha/2}\hat{\sigma}_{\hat{p}}<P<\hat{p}+z_{\alpha/2}\hat{\sigma}_{\hat{p}}$

or

$0.25-(1.96)(0.03155)<P<0.25+(1.96)(0.03155)$

or

(0.1882, 0.3118)

b. As stated, $N=854$, $n=81$, and $x=50$. For a 99% confidence level, note that $z_{\alpha/2}=z_{0.005}=2.576$. Next calculate $\hat{p}$ and $\hat{\sigma}_{\hat{p}}$.

$$\hat{p}=\frac{x}{n}=\frac{50}{80}=0.6173$$

$$\hat{\sigma}_{\hat{p}}^2=\frac{\hat{p}(1-\hat{p})}{n-1}\times\frac{(N-n)}{(N-1)}=\frac{0.6173(1-0.6173)}{81-1}\times\frac{854-81}{854-1}=0.002643$$

$\hat{\sigma}_{\hat{p}}=\sqrt{0.002643}=0.05141$

So the 99% confidence interval is

$\hat{p}-z_{\alpha/2}\hat{\sigma}_{\hat{p}}<P<\hat{p}+z_{\alpha/2}\hat{\sigma}_{\hat{p}}$

or

$0.6173-(2.576)(0.05141)<P<0.6173-(2.576)(0.05141)$

or

(0.4849, 0.7497)

7.56 a. $\bar{x} = 9.7, s = 6.2$, $\hat{\sigma}_{\bar{x}} = \sqrt{\frac{(s)^2}{n}\frac{N-n}{N}} = \sqrt{\frac{(6.2)^2}{50}\frac{139}{189}} = .7519$

$9.7 \pm 1.96\ (.7519)$ (8.2262, 11.1738)

b. 99% confidence interval:

$N\bar{x} - Z_{\alpha/2}N\hat{\sigma}_{\bar{x}} < N\mu < N\bar{x} + Z_{\alpha/2}N\hat{\sigma}_{\bar{x}}$

where, $N\bar{x} = (189)(9.7) = 1833.30$

$N\hat{\sigma}_{\bar{x}} = \sqrt{\frac{s^2}{n}N(N-n)} = \sqrt{\frac{(6.2)^2}{50}189(189-50)} = 142.1167$

$1833.30 \pm 2.58(142.1167)$

$1466.6390 < N\mu < 2199.9610$

7.58 a. $\hat{\sigma}_{\bar{x}} = \sqrt{\frac{(5.32)^2}{40}\frac{85}{125}} = .6936$

$7.28 \pm 2.58\ (.6936)$ (5.4904, 9.0696)

b. 90% confidence interval:

$N\bar{x} - Z_{\alpha/2}N\hat{\sigma}_{\bar{x}} < N\mu < N\bar{x} + Z_{\alpha/2}N\hat{\sigma}_{\bar{x}}$, where $N\bar{x} = (125)(7.28) = 910$

$N\hat{\sigma}_{\bar{x}} = \sqrt{\frac{s^2}{n}N(N-n)} = \sqrt{\frac{(5.32)^2}{40}125(125-40)} = 86.7054$

$910 \pm 1.645(86.7054)$

$767.3696 < N\mu < 1{,}052.6304$

7.60 $\bar{x} = 143/35 = 4.0857$

90% confidence interval:

$N\bar{x} - Z_{\alpha/2}N\hat{\sigma}_{\bar{x}} < N\mu < N\bar{x} + Z_{\alpha/2}N\hat{\sigma}_{\bar{x}}$, where $N\bar{x} = (120)(4.0857) = 490.2857$

$N\hat{\sigma}_{\bar{x}} = \sqrt{\frac{s^2}{n}N(N-n)} = \sqrt{\frac{(3.1)^2}{35}120(120-35)} = 52.9210$

$490.2857 \pm 1.645(52.9210)$

$403.2307 < N\mu < 577.3407$

7.62 $\hat{p} = 56/100 = .56$

$\sigma_{\hat{p}} = \sqrt{[\hat{p}(1-\hat{p})/(n-1)][(N-n)/N]}$

$= \sqrt{[(.56)(.44)/99][(420-100)/420]} = .0435$

90% confidence interval: $.56 \pm 1.645(.0435)$: .4884 up to .6316

7.64 $\hat{p} = 31/80 = .3875$

$\sigma_{\hat{p}} = \sqrt{[\hat{p}(1-\hat{p})/(n-1)][(N-n)/N]}$

$= \sqrt{[(.3875)(.6125)/(79)][(420-80)/420]} = .0493$

90% confidence interval: .3875 ± 1.645(.0493): .3064 up to .4686

128.688 < Np < 196.812 or between 129 and 197 students intend to take the final.

7.66 $n = 16, \quad \bar{x} = 150, \quad s = 12, \quad t_{15,.025} = 2.131$

$\bar{x} \pm t_{\alpha/2}\left(s/\sqrt{n}\right)$ = 150 ± 2.131(12/4) = 143.607 up to 156.393

It is recommended that he stock 157 gallons.

7.68 Results from Minitab:

Descriptive Statistics: Passengers7_68

Variable	N	Mean	Median	TrMean	StDev	SE Mean
Passenge	50	136.22	141.00	136.75	24.44	3.46

Variable	Minimum	Maximum	Q1	Q3
Passenge	86.00	180.00	118.50	152.00

One-Sample T: Passengers7_68

Variable	N	Mean	StDev	SE Mean	95.0% CI
Passengers8_	50	136.22	24.44	3.46	(129.27, 143.17)

95% confidence interval: 129.27 up to 143.17

7.70 a. The minimum variance unbiased point estimator of the population mean is the sample mean: $\bar{X} = \frac{\sum X_i}{n} = \frac{27}{8} = 3.375$. The unbiased point estimate of the variance: $s^2 = \frac{\sum x_i^2 - n\bar{x}^2}{n-1} = \frac{94.62 - 8(3.375)^2}{7} = .4993$

b. $\hat{p} = \frac{x}{n} = \frac{3}{8} = .375$

7.72 $n = 174, \quad \bar{x} = 6.06, \quad s = 1.43$

$6.16 - 6.06 = .1 = z_{\alpha/2}(1.43/\sqrt{174}), \quad z_{\alpha/2} = .922$

$\alpha = 2[1 - F_z(.92)] = .3576$

100(1-.3576)% = 64.24%

7.74 $n = 25$ patient records – the average length of stay is 6 days with a standard deviation of 1.8 days

a. find the reliability factor for a 95% interval estimate $t_{\alpha/2} = 2.064$

b. Find the LCL for a 99% confidence interval estimate of the population mean

$ME = t_{\alpha/2}\, s/\sqrt{n} = 6 - 2.064\left(1.8/\sqrt{25}\right)$. The LCL = 5.257

7.76 a. 90% confidence interval reliability factor = $t_{\alpha/2} = 1.729$

b. Find the LCL for a 99% confidence interval

$$LCL = 60.75 - 2.861\, {21.83159}/{\sqrt{20}} = 46.78$$ or approximately 47 passengers.

7.78 n = 100 students at a small university.

a. Estimate the population grade point average with 95% confidence level

One-Sample T: GPA

```
Variable    N     Mean    StDev  SE Mean        95% CI
GPA       100  3.12800  0.36184  0.03618  (3.05620, 3.19980)
```

b. Estimate the population proportion of students who were very dissatisfied (code 1) or moderately dissatisfied (code 2) with parking. Use a 90% confidence level.

Tally for Discrete Variables: Parking

```
Parking  Count  Percent
      1     19    19.00
      2     26    26.00
      3     18    18.00
      4     18    18.00
      5     19    19.00
     N=    100
```

$n = 100, \quad \hat{p} = 45/100 = .45, \quad z_{.05} = 1.645, \ \hat{p} \pm z_{\alpha/2}\sqrt{\frac{\hat{p}(1-\hat{p})}{n}}$

$$.45 \pm 1.645\sqrt{\frac{.45(1-.45)}{100}} = .45 \pm .08184 = .368162 \text{ up to } .53184$$

c. Estimate the proportion of students who were at least moderately satisfied (codes 4 and 5) with on-campus food service

Tally for Discrete Variables: Dining

```
Dining  Count  Percent
     1     14    14.00
     2     26    26.00
     3     21    21.00
     4     20    20.00
     5     19    19.00
    N=    100
```

$n = 100, \quad \hat{p} = 39/100 = .39, \quad z_{.05} = 1.645$

$\hat{p} \pm z_{\alpha/2}\sqrt{\frac{\hat{p}(1-\hat{p})}{n}} = .39 \pm 1.645\sqrt{\frac{.39(1-.39)}{100}} = .39 \pm .08023 = .30977$ up to .47023.

7.80 n = 500 motor vehicle registrations, 200 were mailed, 160 paid in person, remainder paid on-line.

a. Estimate the population proportion to pay for vehicle registration renewals in person, use a 90% confidence level.

Test and CI for One Proportion

```
Test of p = 0.5 vs p not = 0.5
Sample     X     N  Sample p               90% CI
1        160   500  0.320000   (0.285686, 0.354314)
```

The 90% confidence interval is from 28.56856% up to 35.4314%

b. Estimate the population proportion of on-line renewals, use 95% confidence.

Test and CI for One Proportion

```
Test of p = 0.5 vs p not = 0.5
Sample     X     N  Sample p               95% CI
1        140   500  0.280000   (0.240644, 0.319356)
```

The 95% confidence interval is from 24.0644% up to 31.9356%

7.82 From the data in 7.80, find the confidence level if the interval extends from 23.7% up to 32.3%. ME = ½ the width of the confidence interval. .323 – .237 = .086 / 2 = .043 and $\hat{p} = .28$

$$ME = z_{\alpha/2}\sqrt{\frac{\hat{p}(1-\hat{p})}{n}} \quad .043 = z_{\alpha/2}\sqrt{\frac{.28(1-.28)}{500}} \quad \text{solving for z:} \quad z_{\alpha/2} = 2.14$$

Area from the z-table = .4838 x 2 = .9676. The confidence level is 96.76%

7.84 Compute the 98% confidence interval of the mean age of on-line renewal users. n= 460, sample mean = 42.6, s = 5.4.

$$\bar{x} \pm t_{\alpha/2} \, {}^{s}\!/\!{\sqrt{n}} = 42.6 \pm 2.33\left({}^{5.4}\!/\!{\sqrt{460}} \right) = 42.6 \pm .58664 = 42.0134 \text{ up to } 43.18664$$

7.86 a. $\hat{\sigma}^2_{\bar{x}} = \frac{(149.92)^2}{50}\frac{272-50}{272} = 366.888$

99% confidence interval: $492.36 \pm 2.58\sqrt{366.888}$:
442.9139 up to 541.7501

b. 95% confidence interval: $492.36 \pm 1.96\sqrt{366.888}$:
454.8175 up to 529.9025

c. The 90% interval is narrower; the z-score would decline to 1.645

Chapter 8:

Estimation: Additional Topics

8.2 Two normally distributed populations based on dependent samples of n=5 observations

a. Find the margin of error for 90% confidence level

Paired T-Test and CI: Before_Ex8.2, After_Ex8.2

```
Paired T for Before_Ex8.2 - After_Ex8.2
              N       Mean     StDev   SE Mean
Before_Ex8.2  5     8.4000    2.6077    1.1662
After_Ex8.2   5    10.2000    3.1145    1.3928
Difference    5   -1.80000   0.83666   0.37417

90% CI for mean difference: (-2.59766, -1.00234)
```

$$ME = t_{n-1,\alpha/2}\frac{s_d}{\sqrt{n_d}} = ME = 2.132\frac{.83666}{\sqrt{5}} = .79772$$

b. Find the UCL and LCL for a 90% confidence interval

UCL = -2.59766, LCL = -1.00234

c. Find the width of a 95% confidence interval

$$width = 2\left[ME = t_{n-1,\alpha/2}\frac{s_d}{\sqrt{n_d}}\right] = 2\left[2.776\frac{.83666}{\sqrt{5}}\right] = 2.07737$$

8.4 Let X = Without Passive Solar; Y = With Passive Solar; $d_i = x_i - y_i$

$n = 10,\quad \sum d_i = 373,\quad \bar{d} = 37.3,\quad t_{9,.05} = 1.833$

$$s_d = \sqrt{\frac{\sum\left(d_i - \bar{d}\right)^2}{n_d - 1}} = \sqrt{2806.1/9} = 17.6575$$

$37.3 \pm 1.833(17.6575)/\sqrt{10}$

$27.0649 < \mu_x - \mu_y < 47.5351$

8.6 Independent random samples, find a 90% confidence interval estimate of the difference in the means of the two populations

$$(\bar{x} - \bar{y}) \pm z_{\alpha/2}\sqrt{\frac{\sigma^2{}_x}{n_x} + \frac{\sigma^2{}_y}{n_y}} = (400 - 360) \pm 1.645\sqrt{\frac{20^2}{64} + \frac{25^2}{36}} = 40 \pm 7.9933,$$

32.0067 up to 47.9933

8.8 Assuming equal population variances, determine the number of degrees of freedom for each

a. degrees of freedom = $n_x + n_y - 2 = 12 + 14 - 2 = 24$

b. degrees of freedom = $n_x + n_y - 2 = 6 + 7 - 2 = 11$

c. degrees of freedom = $n_x + n_y - 2 = 9 + 12 - 2 = 19$

8.10 Assuming unequal population variances, determine the number of degrees of freedom

a. $$v = \frac{\left[\left(\frac{s_1^2}{n_1}\right)+\left(\frac{s_2^2}{n_2}\right)\right]^2}{\left(\frac{s_1^2}{n_1}\right)^2/(n_1-1)+\left(\frac{s_2^2}{n_2}\right)^2/(n_2-1)} = v = \frac{\left[\left(\frac{6}{12}\right)+\left(\frac{10}{14}\right)\right]^2}{\left(\frac{6}{12}\right)^2/(12-1)+\left(\frac{10}{14}\right)^2/(14-1)} \approx 24$$

b. $$v = \frac{\left[\left(\frac{s_1^2}{n_1}\right)+\left(\frac{s_2^2}{n_2}\right)\right]^2}{\left(\frac{s_1^2}{n_1}\right)^2/(n_1-1)+\left(\frac{s_2^2}{n_2}\right)^2/(n_2-1)} - v = \frac{\left[\left(\frac{30}{6}\right)+\left(\frac{36}{10}\right)\right]^2}{\left(\frac{30}{6}\right)^2/(6-1)+\left(\frac{36}{10}\right)^2/(10-1)} \approx 11$$

c. $$v = \frac{\left[\left(\frac{s_1^2}{n_1}\right)+\left(\frac{s_2^2}{n_2}\right)\right]^2}{\left(\frac{s_1^2}{n_1}\right)^2/(n_1-1)+\left(\frac{s_2^2}{n_2}\right)^2/(n_2-1)} - v = \frac{\left[\left(\frac{16}{9}\right)+\left(\frac{25}{12}\right)\right]^2}{\left(\frac{16}{9}\right)^2/(9-1)+\left(\frac{25}{12}\right)^2/(12-1)} \approx 19$$

d. $$v = \frac{\left[\left(\frac{s_1^2}{n_1}\right)+\left(\frac{s_2^2}{n_2}\right)\right]^2}{\left(\frac{s_1^2}{n_1}\right)^2/(n_1-1)+\left(\frac{s_2^2}{n_2}\right)^2/(n_2-1)} - v = \frac{\left[\left(\frac{30}{6}\right)+\left(\frac{36}{7}\right)\right]^2}{\left(\frac{30}{6}\right)^2/(6-1)+\left(\frac{36}{7}\right)^2/(7-1)} \approx 11$$

8.12 Let X = machine A and Y = machine B.

$\bar{x} = 130,\ \sigma_x = 8.4,\ n_x = 40 \quad \bar{y} = 120,\ \sigma_y = 11.3,\ n_y = 36$

Find the 95% confidence interval for the difference in means

$$(\bar{x}-\bar{y}) \pm z_{\alpha/2}\sqrt{\frac{\sigma_x^2}{n_x}+\frac{\sigma_y^2}{n_y}} = (130-120) \pm 1.96\sqrt{\frac{8.4^2}{40}+\frac{11.3^2}{36}} = 10 \pm$$

4.5169, 5.4831 up to 14.5169

8.14

Descriptive Statistics: Machine 1, Machine 2

```
Variable            N       Mean     Median     TrMean      StDev    SE Mean
Machine           100     520.95     518.75     520.52       9.45       0.95
Machine           100     513.75     514.05     513.91       5.49       0.55

Variable      Minimum    Maximum         Q1         Q3
Machine        504.70     544.80     513.80     527.28
Machine        496.50     527.00     510.33     517.68
```

95% confidence level: assuming normal populations and similar variances

$$(520.95-513.75)\pm(1.96)\sqrt{\frac{(9.45)^2}{100}+\frac{(5.49)^2}{100}} = 5.0579 \text{ up to } 9.3421$$

8.16 $n_1 = 200, \quad \bar{x} = .517, \quad s_1 = .148, \quad z_{.005} = 2.58$

$n_2 = 400, \quad \bar{y} = .489, \quad s_2 = .159$

$$(.517-.489)\pm(2.58)\sqrt{\frac{(.148)^2}{200}+\frac{(.159)^2}{400}} = -.00591 \text{ up to } .061907$$

8.18 Calculate the margin of error, assuming 95% confidence level

a. $ME = z_{\alpha/2}\sqrt{\frac{\hat{p}_1(1-\hat{p}_1)}{n_1}+\frac{\hat{p}_2(1-\hat{p}_2)}{n_2}} = 1.96\sqrt{\frac{.75(1-.75)}{260}+\frac{.68(1-.68)}{200}} = .083367$

b. $ME = z_{\alpha/2}\sqrt{\frac{\hat{p}_1(1-\hat{p}_1)}{n_1}+\frac{\hat{p}_2(1-\hat{p}_2)}{n_2}} = 1.96\sqrt{\frac{.60(1-.60)}{400}+\frac{.68(1-.68)}{500}} = .063062$

c. $ME = z_{\alpha/2}\sqrt{\frac{\hat{p}_1(1-\hat{p}_1)}{n_1}+\frac{\hat{p}_2(1-\hat{p}_2)}{n_2}} = 1.96\sqrt{\frac{.20(1-.20)}{500}+\frac{.25(1-.25)}{375}} = .056126$

8.20 $n_x = 120, \quad \hat{p}_x = \frac{x}{n} = \frac{85}{120} = .7083, \quad n_y = 163, \quad \hat{p}_y = \frac{y}{n} = \frac{78}{163} = .4785, \quad z_{.01} = 2.33$

$$(\hat{p}_x - \hat{p}_y) \pm z_{\alpha/2}\sqrt{\frac{\hat{p}_x(1-\hat{p}_x)}{n_x}+\frac{\hat{p}_y(1-\hat{p}_y)}{n_y}} =$$

$$(.7083-.4785)\pm(2.326)\sqrt{\frac{(.7083)(.2917)}{120}+\frac{(.4785)(.5215)}{163}} =$$

$.2298 \pm .132657 = .0971$ up to $.3625$

8.22 $\hat{p}_{freshmen} = 80/138 = .5797, \quad \hat{p}_{sophs} = 73/96 = .7604$

$$(.5797-.7604)\pm(1.96)\sqrt{\frac{(.5797)(.4203)}{138}+\frac{(.7604)(.2396)}{96}} =$$

$-.1807 \pm .1187 = -.3001$ up to $-.0627$

8.24 $n_x = 510,\ \hat{p}_x = .6275,\ n_y = 332,\ \hat{p}_y = .6024,\ z_{.05} = 1.645$

$$(.6275-.6024) \pm (1.645)\sqrt{\frac{(.6275)(.3725)}{510}+\frac{(.6024)(.3976)}{332}}$$

$.0251 \pm .0565 = -.0314$ up to $.0816$

8.26 How large a sample is needed to estimate the population proportion?

a. $n = \dfrac{.25(z_{\alpha/2})^2}{ME^2} = \dfrac{.25(1.96)^2}{.03^2} = 1067.111$. Take a sample of size n = 1068.

b. $n = \dfrac{.25(z_{\alpha/2})^2}{ME^2} = \dfrac{.25(1.96)^2}{.05^2} = 384.16$. Take a sample of size n = 385.

c. In order to reduce the ME in half, the sample size must be increased by a larger proportion

8.28 a. $z_{.05} = 1.645,\ ME = .04$

$$n = \frac{.25(z_{\alpha/2})^2}{ME^2} = \frac{(.25)(1.645)^2}{(.04)^2} = 422.8\text{, take n = 423}$$

b. $\dfrac{(.25)(1.96)^2}{(.04)^2} = 600.25$, take n = 601

c. $\dfrac{(.25)(2.33)^2}{(.05)^2} = 542.89$, take n = 543

8.30 $z_{.05} = 1.645,\ B = .03$

$$n = \frac{.25(z_{\alpha/2})^2}{ME^2} = \frac{(.25)(1.645)^2}{(.03)^2} = 751.7\text{, take n = 752}$$

8.32 Use the equation $n = \dfrac{N\sigma^2}{(N-1)\sigma_{\bar{x}}^2+\sigma^2}$ to find the sample size needed.

a. Since $1.96\sigma_{\bar{x}} = 50,\ \sigma_{\bar{x}} = \dfrac{50}{1.96} = 25.51$.

$$n = \frac{N\sigma^2}{(N-1)\sigma_{\bar{x}}^2+\sigma^2} = \frac{(3300)(500)^2}{(3300-1)(25.51)^2+(500)^2} = 344.2\text{, take } n = 345$$

b. As in part (a), $\sigma_{\bar{x}} = 25.51$.

$$n = \frac{N\sigma^2}{(N-1)\sigma_{\bar{x}}^2+\sigma^2} = \frac{(4950)(500)^2}{(4950-1)(25.51)^2+(500)^2} = 356.6\text{, take } n = 357$$

c. As in part (a), $\sigma_{\bar{x}} = 25.51$.

$$n = \frac{N\sigma^2}{(N-1)\sigma_{\bar{x}}^2+\sigma^2} = \frac{(5000000)(500)^2}{(5000000-1)(25.51)^2+(500)^2} = 384.1\text{, take } n = 385$$

d. The required sample size for part (b) is larger than that for part (a), and the required sample size for part (c) is larger than that for parts (a) and (b). This shows that the required sample size increases as N increases.

8.34 $\sigma_{\bar{x}} = \dfrac{2000}{1.96} = 1020.4$, $n = \dfrac{812(20000)^2}{811(1020.4)^2 + (20000)^2} = 261.0038 = 262$ observations

8.36 $\sigma_{\hat{p}} = \dfrac{.05}{2.575} = .0194$

$$n = \frac{.25N}{(N-1)\sigma^2_{p_x} + .25} = \frac{(.25)320}{319(.0194)^2 + .25} = 216.18 = 217 \text{ observations}$$

8.38 Independent random samples from two normally distributed populations. Assuming equal variances, find the 90% confidence interval

$$(\bar{x} - \bar{y}) \pm t_{n_1+n_2-2,\alpha/2} s_p \sqrt{\frac{1}{n_x} + \frac{1}{n_y}} \text{ where } s_p = \sqrt{\frac{(n_x - 1)s^2_x + (n_y - 1)s^2_y}{n_x + n_y - 2}}$$

$$s_p = \sqrt{\frac{(15-1)20^2 + (13-1)25^2}{15+13-2}} = 22.4465$$

$$(400 - 360) \pm 1.706(22.4465)\sqrt{\frac{1}{15} + \frac{1}{13}} = 40 \pm 14.5107 = 25.4893 \text{ to } 54.5107$$

8.40 Independent random samples from two normally distributed populations

a. If the unknown population variances are equal, find a 90% confidence interval

$$(\bar{x} - \bar{y}) \pm t_{n_1+n_2-2,\alpha/2} s_p \sqrt{\frac{1}{n_x} + \frac{1}{n_y}} \text{ where } s_p = \sqrt{\frac{(n_x - 1)s^2_x + (n_y - 1)s^2_y}{n_x + n_y - 2}}$$

$$s_p = \sqrt{\frac{(10-1)30^2 + (12-1)25^2}{10+12-2}} = 27.3633$$

$$(480 - 520) \pm 1.725(27.3633)\sqrt{\frac{1}{10} + \frac{1}{12}} = -40 \pm 20.2106 = -60.21056 \text{ to } -19.7894$$

b. If the unknown population variances are unequal, find a 90% CI

$$(\bar{X}-\bar{Y}) \pm t_{(v,\alpha/2)}\sqrt{\frac{s_x^2}{n_x}+\frac{s_y^2}{n_y}} \text{ where } v=\frac{\left[\left(\frac{s_x^2}{n_x}\right)+\left(\frac{s_y^2}{n_y}\right)\right]^2}{\left(\frac{s_x^2}{n_x}\right)^2/(n_x-1)+\left(\frac{s_y^2}{n_y}\right)^2/(n_y-1)}=$$

$$v=\frac{\left[\left(\frac{30^2}{10}\right)+\left(\frac{25^2}{12}\right)\right]^2}{\left(\frac{30^2}{10}\right)^2/(10-1)+\left(\frac{25^2}{12}\right)^2/(12-1)}=17.606 \approx 18$$

$$(480-520) \pm 1.734\sqrt{\frac{30^2}{10}+\frac{25^2}{12}} = -40 \pm 20.669 = -60.669 \text{ to } -19.331$$

8.42 $n_x = 21$, $\bar{x} = 72.1$, $s_x = 11.3$, $t_{37,.10} = 1.303$ (df = 37 does not appear in Appendix Table 7; we used df = 40 to give an approximate answer)
$n_y = 18$, $\bar{y} = 73.8$, $s_y = 10.6$

$$(\bar{x}-\bar{y}) \pm t_{n_1+n_2-2,\alpha/2} s_p\sqrt{\frac{1}{n_x}+\frac{1}{n_y}} \text{ where } s_p=\sqrt{\frac{(n_x-1)s^2{}_x+(n_y-1)s^2{}_y}{n_x+n_y-2}}$$

$$s_p=\sqrt{\frac{(21-1)11.3^2+(18-1)10.6^2}{21+18-2}}=10.9839$$

$$(72.1-73.8) \pm 1.303(10.9839)\sqrt{\frac{1}{21}+\frac{1}{18}} = -1.7 \pm 4.5971 = -6.2971 \text{ to } 2.8971$$

8.44 $\hat{p}_x - \hat{p}_y = .6222 - .5714 = .0508$

a. Minitab results:

CI for Two Proportions

```
Sample       X      N  Sample p
1          140    225  0.622222
2          120    210  0.571429
Estimate for p(1) - p(2):  0.0507937
95% CI for p(1) - p(2):  (-0.0413643, 0.142952)
```

8.46 Assume both populations are distributed normally with equal variances and a 90% confidence interval.

$n_x = 15, \quad \bar{x} = 470, \quad s_x = 5, \quad t_{25,.05} = 1.708$

$n_y = 12, \quad \bar{y} = 460, \quad s_y = 7$

$$(470-460) \pm (1.708)\sqrt{\frac{(n_x-1)s_x^{\ 2}+(n_y-1)s_y^{\ 2}}{n_x+n_y-2}}\sqrt{\frac{1}{n_x}+\frac{1}{n_y}}$$

$$(470-460) \pm (1.708)\sqrt{\frac{(15-1)5^2+(12-1)7^2}{15+12-2}}\sqrt{\frac{1}{15}+\frac{1}{12}} \quad 10 \pm (1.708)(5.9632)(.3873)$$

$= 10 \pm 3.9447 = 6.055$ up to 13.945

Since both endpoints of the confidence interval are positive, this provides evidence that the new machine provides a larger mean filling weight than the old

8.48 98% confidence interval for student pair:

Paired T-Test and CI: COURSE, NO COURSE

```
Paired T for COURSE - NO COURSE
              N      Mean     StDev   SE Mean
COURSE        6   70.6667   16.0333    6.5456
NO COURSE     6   66.1667   14.1904    5.7932
Difference    6   4.50000   4.13521   1.68819

98% CI for mean difference: (-1.18066, 10.18066)
```

8.50 Construct a 95% confidence interval of the difference in population proportions

$$\hat{p}_1 = x/n = 300/400 = .75, \ \hat{p}_2 = x/n = 225/500 = .45$$

$$(\hat{p}_1 - \hat{p}_2) \pm z_{\alpha/2}\sqrt{\frac{\hat{p}_1(1-\hat{p}_1)}{n_1}+\frac{\hat{p}_2(1-\hat{p}_2)}{n_2}} =$$

$$(.75-.45) \pm 1.96\sqrt{\frac{.75(1-.75)}{400}+\frac{.45(1-.45)}{500}} = .30 \pm .06085 = .23915 \text{ up to } .36085$$

8.52 $\sigma_{\hat{p}} = \dfrac{.06}{1.96} = .0306, \ n = \dfrac{527(.25)}{526(.0306)^2+.25} = 177.43 = 178$ observations

Chapter 9:

Hypothesis Testing: Single Population

9.2 H_0: No change in interest rates is warranted
H_1: Reduce interest rates to stimulate the economy

9.4 a. European perspective:
H_0: Genetically modified food stuffs are not safe
H_1: They are safe
b. U.S. farmer perspective:
H_0: Genetically modified food stuffs are safe
H_1: They are not safe

9.6 A random sample is obtained from a population with a variance of 625 and the sample mean is computed. Test the null hypothesis $H_0: \mu = 100$ versus the alternative $H_1: \mu \geq 100$. Compute the critical value $\bar{x}_c$ and state the decision rule

a. n = 25. Reject H_0 if $\bar{x} > \bar{x}_c = \mu_0 + z_\alpha \sigma/\sqrt{n}$ = 100 +1.645(25)/ $\sqrt{25}$ = 108.225
b. n = 16. Reject H_0 if $\bar{x} > \bar{x}_c = \mu_0 + z_\alpha \sigma/\sqrt{n}$ = 100 +1.645(25)/ $\sqrt{16}$ = 110.28125
c. n = 44. Reject H_0 if $\bar{x} > \bar{x}_c = \mu_0 + z_\alpha \sigma/\sqrt{n}$ = 100 +1.645(25)/ $\sqrt{44}$ = 106.1998
d. n = 32 Reject H_0 if $\bar{x} > \bar{x}_c = \mu_0 + z_\alpha \sigma/\sqrt{n}$ = 100 +1.645(25)/ $\sqrt{32}$ = 107.26994

9.8 Using the results from the above two exercises, indicate how the critical value $\bar{x}_c$ is influenced by sample size. Next indicate how the critical value is influenced by the population variance.
The critical value $\bar{x}_c$ is farther away from the hypothesized value the smaller the sample size n. This is due to the increase in the standard error with a smaller sample size.
The critical value $\bar{x}_c$ is farther away from the hypothesized value the larger the population variance. This is due to the increased standard error with a larger population variance.

9.10 A random sample of n = 25, variance = σ^2 and the sample mean is = 70. Consider the null hypothesis $H_0: \mu = 80$ versus the alternative $H_1: \mu \le 80$. Compute the p-value

a. $\sigma^2 = 225$. $z = \frac{\bar{x} - \mu_0}{\sigma/\sqrt{n}} = \frac{70-80}{15/\sqrt{25}}$ = -3.33. $p-value = P(z_p < -3.33)$ = .0004

b. $\sigma^2 = 900$. $z = \frac{\bar{x} - \mu_0}{\sigma/\sqrt{n}} = \frac{70-80}{30/\sqrt{25}}$ = -1.67. $p-value = P(z_p < -1.67)$ = .0475

c. $\sigma^2 = 400$. $z = \frac{\bar{x} - \mu_0}{\sigma/\sqrt{n}} = \frac{70-80}{20/\sqrt{25}}$ = -2.50. $p-value = P(z_p < -2.50)$ = .0062

d. $\sigma^2 = 600$. $z = \frac{\bar{x} - \mu_0}{\sigma/\sqrt{n}} = \frac{70-80}{24.4949/\sqrt{25}}$ = -2.04. $p-value = P(z_p < -2.04)$ = .0207

9.12 $H_0: \mu \ge 50$; $H_1: \mu < 50$; reject H_0 if $Z_{.10} < -1.28$

$Z = \frac{48.2-50}{3/\sqrt{9}}$ = -1.8, therefore, Reject H_0 at the 10% level.

9.14 Test $H_0: \mu \le 100$; $H_1: \mu > 100$, using n = 25 and alpha = .05

a. $\bar{x} = 106, s = 15$. Reject if $\frac{\bar{x} - \mu_0}{s/\sqrt{n}} > t_{n-1,\alpha/2}$, $\frac{106-100}{15/\sqrt{25}}$ = 2.00. Since 2.00 is greater than the critical value of 1.711, there is sufficient evidence to reject the null hypothesis.

b. $\bar{x} = 104, s = 10$. Reject if $\frac{\bar{x} - \mu_0}{s/\sqrt{n}} > t_{n-1,\alpha/2}$, $\frac{104-100}{10/\sqrt{25}}$ = 2.00. Since 2.00 is greater than the critical value of 1.711, there is sufficient evidence to reject the null hypothesis.

c. Assuming a one-tailed lower tailed test, $\bar{x} = 95, s = 10$. Reject if $\frac{\bar{x} - \mu_0}{s/\sqrt{n}} > t_{n-1,\alpha/2}$, $\frac{95-100}{10/\sqrt{25}}$ = -2.50. Since -2.50 is less than the critical value of -1.711, there is sufficient evidence to reject the null hypothesis.

d. Assuming a one-tailed lower test, $\bar{x} = 92, s = 18$. Reject if $\frac{\bar{x} - \mu_0}{s/\sqrt{n}} > t_{n-1,\alpha/2}$, $\frac{92-100}{18/\sqrt{25}}$ = -2.22. Since -2.22 is less than the critical value of -1.711, there is sufficient evidence to reject the null hypothesis.

9.16 $H_0: \mu \geq 3; H_1: \mu < 3;$

$t = \frac{2.4-3}{1.8/\sqrt{100}}$ = -3.33, p-value is < .005. Reject H_0 at any common level of alpha

9.18 $H_0: \mu = 0; H_1: \mu \neq 0;$

$t = \frac{.078-0}{.201/\sqrt{76}}$ = 3.38, p-value is < .010. Reject H_0 at any common level of alpha

9.20 $H_0: \mu = 0; H_1: \mu < 0;$

$t = \frac{-2.91-0}{11.33/\sqrt{170}}$ = -3.35, p-value is < .005. Reject H_0 at any common level of alpha.

9.22 a. No, the 95% confidence level provides for 2.5% of the area in either tail. This does not correspond to a one-tailed hypothesis test with an alpha of 5% which has 5% of the area in one of the tails.
b. Yes.

9.24 $H_0: \mu = 20; H_1: \mu \neq 20;$ reject H_0 if $|t_{8,.05/2}| > 2.306$

$t = \frac{20.3556-20}{.6126/\sqrt{9}}$ = 1.741, therefore, do not reject H_0 at the 5% level

9.26 The population values must be assumed to be normally distributed.
$H_0: \mu \geq 50; H_1: \mu < 50;$ reject H_0 if $t_{19,.05} < -1.729$

$t = \frac{41.3-50}{12.2/\sqrt{20}}$ = -3.189, therefore, reject H_0 at the 5% level

9.28 A random sample is obtained to test the null hypothesis of the proportion of women who said yes to a new shoe model. $H_0: p \leq .25; H_1: p > .25;$. What value of the sample proportion is required to reject the null hypothesis with alpha = .03?

a. n = 400. Reject H_0 if $\hat{p} > \hat{p}_c = p_0 + z_\alpha\sqrt{p_0(1-p_0)/n}$ = .25 +1.88 $\sqrt{(.25)(1-.25)/400}$ = .2907

b. n = 225. Reject H_0 if $\hat{p} > \hat{p}_c = p_0 + z_\alpha\sqrt{p_0(1-p_0)/n}$ = .25 +1.88 $\sqrt{(.25)(1-.25)/225}$ = .30427

c. n = 625. Reject H_0 if $\hat{p} > \hat{p}_c = p_0 + z_\alpha\sqrt{p_0(1-p_0)/n}$ = .25 +1.88 $\sqrt{(.25)(1-.25)/625}$ = .28256

d. n = 900. Reject H_0 if $\hat{p} > \hat{p}_c = p_0 + z_\alpha\sqrt{p_0(1-p_0)/n}$ = .25 +1.88 $\sqrt{(.25)(1-.25)/900}$ = .2771

9.30 $H_0: p \leq .25; H_1: p > .25;$

$z = \frac{.2908 - .25}{\sqrt{(.25)(.75)/361}}$ = 1.79, p-value = 1 – F_Z(1.79) = 1 - .9633 = .0367

Therefore, reject H_0 at alpha greater than 3.67%

9.32 $H_0: p = .5; H_1: p \neq .5;$

$z = \frac{.45 - .5}{\sqrt{(.5)(.5)/160}}$ = -1.26, p-value = 2[1 – F_Z(1.26)] = 2[1 – .8962] = .2076

The probability of finding a random sample with a sample proportion this far or further from .5 if the null hypothesis is really true is .2076

9.34 $H_0: p = .5; H_1: p > .5;$

$z = \frac{.56 - .5}{\sqrt{(.5)(.5)/50}}$ = .85, p-value = 1 – F_Z(.85) = 1 – .8023 = .1977

Therefore, reject H_0 at alpha levels in excess of 19.77%

9.36 $H_0: p \geq .75; H_1: p < .75;$

$z = \frac{.6931 - .75}{\sqrt{(.75)(.25)/202}}$ = -1.87, p-value = 1 – F_Z(1.87) = 1 – .9693 = .0307

Therefore, reject H_0 at alpha levels in excess of 3.07%

9.38 What is the probability of Type II error if the actual proportion is

a. $P = .52$. $\beta = P(.46 \leq \hat{p} \leq .54 \mid p = p^*)$

$$= P\left[\frac{.46 - p^*}{\sqrt{\frac{p^*(1-p^*)}{n}}} \leq z \leq \frac{.54 - p^*}{\sqrt{\frac{p^*(1-p^*)}{n}}}\right] = P\left[\frac{.46 - .52}{\sqrt{\frac{.52(1-.52)}{600}}} \leq z \leq \frac{.54 - .52}{\sqrt{\frac{.52(1-.52)}{600}}}\right]$$

= P(-2.94 ≤ z ≤ .98) = .4984 + .3365 = .8349

b. $P = .58$. $\beta = P(.46 \le \hat{p} \le .54 \mid p = p^*)$

$$= P\left[\frac{.46 - .58}{\sqrt{\frac{.58(1-.58)}{600}}} \le z \le \frac{.54 - .58}{\sqrt{\frac{.58(1-.58)}{600}}}\right] = \text{P}(-5.96 \le z \le -1.99) = .5000 - .4767 = .0233$$

c. $P = .53$. $\beta = P(.46 \le \hat{p} \le .54 \mid p = p^*)$

$$= P\left[\frac{.46 - .53}{\sqrt{\frac{.53(1-.53)}{600}}} \le z \le \frac{.54 - .53}{\sqrt{\frac{.53(1-.53)}{600}}}\right] = \text{P}(-3.44 \le z \le .49) = .4997 + .1879 = .6876$$

d. $P = .48$. $\beta = P(.46 \le \hat{p} \le .54 \mid p = p^*)$

$$= P\left[\frac{.46 - .48}{\sqrt{\frac{.48(1-.48)}{600}}} \le z \le \frac{.54 - .48}{\sqrt{\frac{.48(1-.48)}{600}}}\right] = \text{P}(-.98 \le z \le 2.94) = .3365 + .4984 = .8349$$

e. $P = .43$. $\beta = P(.46 \le \hat{p} \le .54 \mid p = p^*)$

$$= P\left[\frac{.46 - .43}{\sqrt{\frac{.43(1-.43)}{600}}} \le z \le \frac{.54 - .43}{\sqrt{\frac{.43(1-.43)}{600}}}\right] = \text{P}(1.48 \le z \le 5.44) = .5000 - .4306 = .0694$$

9.40 a. H_0 is rejected when $\frac{\bar{X} - 3}{.4/\sqrt{64}} > 1.645$ or when $\bar{X} > 3.082$. Since the sample mean is 3.07% which is less than the critical value, the decision is do not reject the null hypothesis.

b. The β = P(Z < $\frac{3.082 - 3.1}{.4/\sqrt{64}}$) = 1 – $F_Z(.36)$ = .3594. Power of the test = 1 - β = .6406

9.42 H_0 is rejected when $\frac{p - .5}{\sqrt{.25/802}} < -1.28$ or when p < .477

The power of the test = 1 - β = 1 – P(Z > $\frac{.477 - .45}{\sqrt{(.45)(.55)/802}}$) = 1-P(Z > 1.54) = .9382

9.44 a. H_0 is rejected when $-1.645 > \frac{p - .5}{\sqrt{.25/199}} > 1.645$ or when .442 > p > .558.
Since the sample proportion is .5226 which is within the critical values. The decision is that there is insufficient evidence to reject the null hypothesis.

b. β = P($\frac{.442 - .6}{\sqrt{(.6)(.4)/199}}$ < Z < $\frac{.558 - .6}{\sqrt{(.6)(.4)/199}}$) = 1-P(-4.55 < Z < -1.21) = .1131

9.46 a. $\alpha = P(Z > \frac{.14-.10}{\sqrt{(.1)(.9)/100}}) = P(Z > 1.33) = .0918$

b. $\alpha = P(Z > \frac{.14-.10}{\sqrt{(.1)(.9)/400}}) = P(Z > 2.67) = .0038$. The smaller probability of a Type I error is due to the larger sample size which lowers the standard error of the mean.

c. $\beta = P(Z < \frac{.14-.20}{\sqrt{(.2)(.8)/100}}) = P(Z < -1.5) = .0668$

d. i) lower; ii) higher

9.48 $H_0: \sigma^2 \le 500; H_1: \sigma^2 > 500$; reject H_0 if $\chi^2_{(7,.10)} > 12.02$

$$\chi^2 = \frac{(n-1)s^2}{\sigma^2} = \frac{7(933.982)}{500} = 13.0757,$$

Therefore, reject H_0 at the 10% level

9.50 $H_0: \sigma^2 = 300; H_1: \sigma^2 \ne 300;$

$\chi^2 = \frac{29(480)}{300} = 46.4$, p-value = .0214. Reject H_0 at the 5% level

9.52 $H_0: \sigma \ge 18.2; H_1: \sigma < 18.2;$

$$\chi^2 = \frac{24(15.3)^2}{(18.2)^2} = 16.961.$$

Do not reject H_0 at the 10% level since $\chi^2 > 15.66 = \chi^2_{(24,.10)}$

9.54 The p-value indicates the likelihood of getting the sample result at least as far away from the hypothesized value as the one that was found, assuming that the distribution is really centered on the null hypothesis. The smaller the p-value, the stronger the evidence against the null hypothesis.

9.56 a. False. The significance level is the probability of making a Type I error – falsely rejecting the null hypothesis when in fact the null is true.

b. True

c. True

d. False. The power of the test is the ability of the test to correctly reject a false null hypothesis.

e. False. The rejection region is farther away from the hypothesized value at the 1% level than it is at the 5% level. Therefore, it is still possible to reject at the 5% level but not at the 1% level.

f. True

g. False. The p-value tells the strength of the evidence against the null hypothesis.

9.58 a. $\alpha = P(Z < \frac{776-800}{120/\sqrt{100}})$ = P(Z < -2) = .0228

b. $\beta = P(Z > \frac{776-740}{120/\sqrt{100}})$ = P(Z > 3) = .0014

c. i) smaller; ii) smaller

d. i) smaller; ii) larger

9.60 $H_0: p = .5; H_1: p \neq .5;$

$z = \frac{.4808-.5}{\sqrt{(.5)(.5)/104}}$ = -.39, p-value = 2[1-F_Z(.39)] = 2[1-.6517] = .6966

Therefore, reject H_0 at levels in excess of 69.66%

9.62 $H_0: p \leq .25; H_1: p > .25;$ reject H_0 if $z_{.05}$ > 1.645

$z = \frac{.3333-.25}{\sqrt{(.25)(.75)/150}}$ = 2.356, therefore, reject H_0 at the 5% level

9.64 Cost Model where W = Total Cost: W = 1,000 + 5X

$\mu_W = 1,000 + 5(400) = 3,000$

$\sigma^2_W = (5)^2(625) = 15,625, \quad \sigma_W = 125, \quad \sigma_{\bar{W}} = \frac{125}{\sqrt{25}} = 25$

H_0: W ≤ 3000; $H1: W > 3000;$

Using the test statistic criteria: (3050 – 3000)/25 = 2.00 which yields a p-value of .0228, therefore, reject H_0 at the .05 level.

Using the sample statistic criteria: $\bar{X}_{crit} = 3,000 + (25)(1.645) = 3041.1$,

$\bar{X}_{calc} = 3,050$, since $\bar{X}_{calc} = 3,050$ > $\bar{X}_{crit} = 3041.1$, therefore, reject H_0 at the .05 level.

9.66 Assume that the population of matched differences are normally distributed

$H_o: \mu_x - \mu_y = 0; H_1: \mu_x - \mu_y \neq 0;$

$t = \frac{1.3667-0}{2.414/\sqrt{12}}$ = 1.961.

Reject H_0 at the 10% level since 1.961 > 1.796 = $t_{(11, .05)}$

9.68 $H_0: \mu \leq 40, H_1: \mu > 40; \bar{X} = 49.73 > 42.86$ reject H_0

One-Sample T: Salmon Weight

```
Test of mu = 40 vs mu > 40

Variable          N       Mean      StDev    SE Mean
Salmon Weigh     39      49.73      10.60       1.70

Variable       95.0% Lower Bound          T        P
Salmon Weigh                46.86      5.73    0.000
```

At the .05 level of significance we have strong enough evidence to reject Ho that the true mean weight of salmon is no different than 40 in favor of Ha that the true mean weight is significantly greater than 40.

$\bar{X}_{crit} = Ho + t_{crit}(S_{\bar{x}})$: 40 + 1.686(1.70) = 42.8662

Population mean for β = .50 (power=.50): tcrit = 0.0: 42.8662 + 0.0(1.70) = 42.8662

Population mean for β = .25 (power=.75): tcrit = .681: 42.8662 + .681(1.70) = 44.0239

Population mean for β = .10 (power=.90): tcrit = 1.28: 42.8662 + 1.28(1.70) = 45.0422

Population mean for β = .05 (power=.95): tcrit = 1.645: 42.8662 + 1.645(1.70) = 45.6627

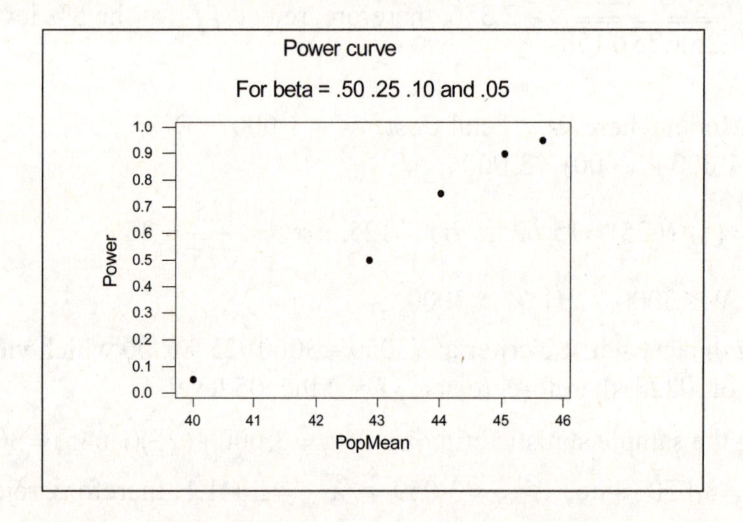

9.70 a. Assume that the population is normally distributed

One-Sample T: Grams:

```
Test of mu = 5 vs mu not = 5
Variable          N       Mean      StDev    SE Mean
Grams:11-34      12     4.9725     0.0936     0.0270

Variable              95.0% CI              T        P
Grams:11-34      (  4.9130,   5.0320)    -1.02    0.331
```

$\bar{x} = 4.9725; s = .0936$, $H_0: \mu = 5; H_1: \mu \neq 5$; reject H_0 if $|t_{(11,.025)}| > 2.201$

$t = \dfrac{4.9725 - 5}{.0936/\sqrt{12}}$ = -1.018. Do not reject H_0 at the 5% level

b. Assume that the population is normally distributed

$H_0: \sigma = .025; H_1: \sigma > .025$; reject H_0 if $\chi^2_{(11,.05)} > 19.68$

$\chi^2 = \dfrac{11(.0936)^2}{(.025)^2}$ = 154.19. Therefore, reject H_0 at the 5% level

Chapter 10:

Hypothesis Testing: Additional Topics

10.2 n = 25 paired observations with sample means of 50 and 56 for populations 1 and 2. Can you reject the null hypothesis at an alpha of .05 if

a. $s_d = 20,\ H_0: \mu_1 - \mu_2 \geq 0; H_1: \mu_1 - \mu_2 < 0;$

$t = \frac{(-6) - 0}{20/\sqrt{25}}$ = -1.50, p-value = .073. Do not reject H_0 at alpha of .05

Paired T-test and CI

```
              N       Mean      StDev  SE Mean
Difference   25   -6.00000   20.00000  4.00000
95% upper bound for mean difference: 0.84353
T-Test of mean difference = 0 (vs < 0): T-Value = -1.50  P-Value = 0.073
```

b. $s_d = 30,\ H_0: \mu_1 - \mu_2 \geq 0; H_1: \mu_1 - \mu_2 < 0;$

$t = \frac{(-6) - 0}{30/\sqrt{25}}$ = -1.00, p-value = .164. Do not reject H_0 at alpha of .05

Paired T-Test and CI

```
              N       Mean      StDev  SE Mean
Difference   25   -6.00000   30.00000  6.00000
95% upper bound for mean difference: 4.26529
T-Test of mean difference = 0 (vs < 0): T-Value = -1.00  P-Value = 0.164
```

c. $s_d = 15,\ H_0: \mu_1 - \mu_2 \geq 0; H_1: \mu_1 - \mu_2 < 0;$

$t = \frac{(-6) - 0}{15/\sqrt{25}}$ = -2.00, p-value = .028. Reject H_0 at alpha of .05

Paired T-Test and CI

```
              N       Mean     StDev  SE Mean
Difference   25   -6.00000  15.00000  3.00000
95% upper bound for mean difference: -0.86735
T-Test of mean difference = 0 (vs < 0): T-Value = -2.00  P-Value = 0.028
```

d. $s_d = 40,\ H_0: \mu_1 - \mu_2 \geq 0; H_1: \mu_1 - \mu_2 < 0;$

$t = \frac{(-6) - 0}{40/\sqrt{25}}$ = -.75, p-value = .230. Do not reject H_0 at alpha of .05

Paired T-Test and CI

```
              N       Mean      StDev  SE Mean
Difference   25   -6.00000   40.00000  8.00000
95% upper bound for mean difference: 7.68706
T-Test of mean difference = 0 (vs < 0): T-Value = -0.75  P-Value = 0.230
```

10.4 $H_0: \mu_x - \mu_y = 0; H_1: \mu_x - \mu_y > 0;$

$$t = \frac{1475 - 0}{1862.985/\sqrt{8}} = 2.239, \text{ p-value} = .0301.$$

Reject H_0 at levels in excess of 3%

Paired T-Test and CI: Male, Female

```
Paired T for Male - Female
             N     Mean     StDev  SE Mean
Male         8  46437.5    2680.1    947.5
Female       8  44962.5    2968.4   1049.5
Difference   8  1475.00   1862.99   658.66
95% lower bound for mean difference: 227.11
T-Test of mean difference = 0 (vs > 0): T-Value = 2.24  P-Value = 0.030
```

10.6 a. Reject H_0 if $\frac{\bar{x} - \bar{y} - D_0}{\sqrt{\frac{\sigma_x^2}{n_x} + \frac{\sigma_y^2}{n_y}}} > z_\alpha$. For $\alpha = 0.05, z_\alpha = z_{0.05} = 1.645$.

$$z = \frac{\bar{x} - \bar{y} - D_0}{\sqrt{\frac{\sigma_x^2}{n_x} + \frac{\sigma_y^2}{n_y}}} = \frac{50 - 60}{\sqrt{\frac{900}{25} + \frac{1600}{28}}} = -1.04$$

Do not reject H_0 at alpha of .05.

b. Reject H_0 if $\frac{\bar{x} - \bar{y} - D_0}{\sqrt{\frac{\sigma_x^2}{n_x} + \frac{\sigma_y^2}{n_y}}} > z_\alpha$. For $\alpha = 0.05, z_\alpha = z_{0.05} = 1.645$.

$$z = \frac{\bar{x} - \bar{y} - D_0}{\sqrt{\frac{\sigma_x^2}{n_x} + \frac{\sigma_y^2}{n_y}}} = \frac{20}{\sqrt{\frac{900}{25} + \frac{1600}{28}}} = 2.07$$

Reject H_0 at alpha of .05.

c. Reject H_0 if $\frac{\bar{x} - \bar{y} - D_0}{\sqrt{\frac{\sigma_x^2}{n_x} + \frac{\sigma_y^2}{n_y}}} > z_\alpha$. For $\alpha = 0.05, z_\alpha = z_{0.05} = 1.645$.

$$z = \frac{\bar{x} - \bar{y} - D_0}{\sqrt{\frac{\sigma_x^2}{n_x} + \frac{\sigma_y^2}{n_y}}} = \frac{45 - 50}{\sqrt{\frac{900}{25} + \frac{1600}{28}}} = -0.52$$

Do not reject H_0 at alpha of .05.

d. Reject H_0 if $\frac{\bar{x}-\bar{y}-D_0}{\sqrt{\frac{\sigma_x^2}{n_x}+\frac{\sigma_y^2}{n_y}}} > z_\alpha$. For $\alpha = 0.05, z_\alpha = z_{0.05} = 1.645$.

$$z = \frac{\bar{x}-\bar{y}-D_0}{\sqrt{\frac{\sigma_x^2}{n_x}+\frac{\sigma_y^2}{n_y}}} = \frac{15}{\sqrt{\frac{900}{25}+\frac{1600}{28}}} = 1.55$$

Do not reject H_0 at alpha of .05.

10.8 $H_0: \mu_x - \mu_y = 0; H_1: \mu_x - \mu_y > 0;$

$$z = \frac{85.8-71.5}{\sqrt{(19.13)^2/151+(12.2)^2/108}} = 7.334.$$

Reject H_0 at all common levels of alpha

10.10 $H_0: \mu_x - \mu_y = 0; H_1: \mu_x - \mu_y \neq 0;$

$$z = \frac{2.71-2.79}{\sqrt{(.64)^2/114+(.56)^2/123}} = \text{-}1.0207,$$

p-value = $2[1\text{-}F_Z(1.02)] = 2[1\text{-}.8461] = .3078$
Therefore, reject H_0 at levels of alpha in excess of 30.78%

10.12 Assuming both populations are normal with equal variances:
$H_0: \mu_x - \mu_y = 0; H_1: \mu_x - \mu_y \neq 0;$

$$s_p^2 = \frac{69(6.14)^2 + 50(4.29)^2}{70+51-2} = 29.592247$$

$$t = \frac{\bar{X}-\bar{Y}-D_0}{\sqrt{\frac{s_p^2}{n_x}+\frac{s_p^2}{n_y}}} = \frac{3.97-2.86}{\sqrt{\frac{29.592247}{70}+\frac{29.592247}{51}}} = 1.108$$

Therefore, do not reject H_0 at the 10% alpha level since $1.108 < 1.645 = t_{(119,.05)}$

10.14 a. $H_0: P_x - P_y = 0; H_1: P_x - P_y < 0;$

$$\hat{p}_o = \frac{500(.42)+600(.50)}{500+600} = .4636,$$

$$z = \frac{.42-.50}{\sqrt{\frac{(.4636)(1-.4636)}{500}+\frac{(.4636)(1-.4636)}{600}}} = \text{-}2.65 \quad \text{p-value} = .004.$$

Therefore, reject H_0 at all common levels of alpha

10.18 $H_0: P_x - P_y = 0; H_1: P_x - P_y \neq 0$; reject H_0 if $|z_{.025}| > 1.96$

$$\hat{p}_o = \frac{78+208}{175+604} = .36714$$

$$z = \frac{.446 - .344}{\sqrt{\frac{(.36714)(.63286)}{175} + \frac{(.36714)(.63286)}{604}}} = 2.465.$$ Reject H_0 at the 5% level

10.20 $H_0: P_x - P_y = 0; H_1: P_x - P_y > 0$; reject H_0 if $|z_{.05}| > 1.645$

$$\hat{p}_o = \frac{138+128}{240+240} = .554$$

$$z = \frac{.575 - .533}{\sqrt{\frac{(.554)(.446)}{240} + \frac{(.554)(.446)}{240}}} = .926.$$

Do not reject H_0 at the 5% level

10.22 a. $H_0: \sigma^2_x = \sigma^2_y; H_1: \sigma^2_x > \sigma^2_y$

F = 125/51 = 2.451. Reject H_0 at the 1% level since 2.451 > 2.11 ≈ $F_{(44,40,.01)}$

b. $H_0: \sigma^2_x = \sigma^2_y; H_1: \sigma^2_x > \sigma^2_y$

F = 235/125 = 1.88. Reject H_0 at the 5% level since 1.88 > 1.69 ≈ $F_{(43,44,.05)}$

c. $H_0: \sigma^2_x = \sigma^2_y; H_1: \sigma^2_x > \sigma^2_y$

F = 134/51 = 2.627. Reject H_0 at the 1% level since 2.627 > 2.11 ≈ $F_{(47,40,.01)}$

d. $H_0: \sigma^2_x = \sigma^2_y; H_1: \sigma^2_x > \sigma^2_y$

F = 167/88 = 1.90. Reject H_0 at the 5% level since 1.90 > 1.79 ≈ $F_{(24,38,.05)}$

10.24 $H_0: \sigma^2_x = \sigma^2_y; H_1: \sigma^2_x > \sigma^2_y$; reject H_0 if $F_{(3,6,.05)} > 4.76$

F = 114.09/16.08 = 7.095. Reject H_0 at the 5% level

10.26 $H_0: \sigma^2_x = \sigma^2_y; H_1: \sigma^2_x \neq \sigma^2_y$;

$F = (2107)^2/(1681)^2 = 1.57$

Therefore, do not reject H_0 at the 10% level since 1.57 < 3.18 ≈ $F_{(9,9,.05)}$

10.28 No. The probability of rejecting the null hypothesis given that it is true is 5%.

10.30 a. $H_0: \mu \le 4; H_1: \mu > 4$; reject H_0 if $t_{.05} > 1.671$

$$t = \frac{4.4 - 4}{1.3/\sqrt{70}} = 2.574.$$ Reject H_0 at the 5% level

b. $H_0: \mu_x - \mu_y = 0; H_1: \mu_x - \mu_y < 0$; reject H_0 if $t_{.05} < -1.645$

$$s_p^2 = \frac{(n_x - 1)s_x^2 + (n_y - 1)s_y^2}{n_x + n_y - 2} = \frac{69(1.3)^2 + 105(1.4)^2}{70 + 106 - 2} = 1.853$$

$$t = \frac{\bar{X} - \bar{Y} - D_0}{\sqrt{\frac{s_p^2}{n_x} + \frac{s_p^2}{n_y}}} = \frac{4.4 - 5.3}{\sqrt{\frac{1.853}{70} + \frac{1.853}{106}}} = -4.293.$$

Reject H_0 at levels in excess of 5%

10.32 Presuming the populations are normally distributed with equal variances, the samples must be independent random samples:

$H_0: \mu_x - \mu_y = 0; H_1: \mu_x - \mu_y < 0$; reject H_0 if $t_{(6,.01)} < -3.143$

$$s_p^2 = \frac{(n_x - 1)s_x^2 + (n_y - 1)s_y^2}{n_x + n_y - 2} = \frac{3(24.4)^2 + 3(14.6)^2}{4 + 4 - 2} = 106.58$$

$$t = \frac{\bar{X} - \bar{Y} - D_0}{\sqrt{\frac{s_p^2}{n_x} + \frac{s_p^2}{n_y}}} = \frac{78 - 114.7}{\sqrt{\frac{106.58}{4} + \frac{106.58}{4}}} = -5.027.$$

Reject H_0 at levels in excess of 1%

10.34 Assume that the populations are normally distributed with equal variances and independent random samples:

Magazine A: $\bar{X} = 7.045; s_x = 2.1819$, Magazine B: $\bar{Y} = 6.777; s_y = 2.85$

$H_0: \mu_x - \mu_y = 0; H_1: \mu_x - \mu_y \ne 0;$

$$s_p^2 = \frac{(n_x - 1)s_x^2 + (n_y - 1)s_y^2}{n_x + n_y - 2} = \frac{5(2.1819)^2 + 5(2.85)^2}{6 + 6 - 2} = 6.4416$$

$$t = \frac{\bar{X} - \bar{Y} - D_0}{\sqrt{\frac{s_p^2}{n_x} + \frac{s_p^2}{n_y}}} = \frac{7.045 - 6.777}{\sqrt{\frac{6.4416}{6} + \frac{6.4416}{6}}} = .183.$$ Do not reject H_0 at any common level of alpha

10.36 $H_0: \mu_x - \mu_y = 0; H_1: \mu_x - \mu_y \neq 0;$. Sample sizes less than 100, use the t-test

$$s_p^2 = \frac{(n_x - 1)s_x^2 + (n_y - 1)s_y^2}{n_x + n_y - 2} = \frac{82(.649)^2 + 53(.425)^2}{83 + 54 - 2} = .32675$$

$$t = \frac{\overline{X} - \overline{Y} - D_0}{\sqrt{\frac{s_p^2}{n_x} + \frac{s_p^2}{n_y}}} = \frac{6.543 - 6.733}{\sqrt{\frac{.32675}{83} + \frac{.32675}{54}}} = -1.901.$$

p-value is between (.05 and .025) x 2 = .10 and .05.
Reject H_0 at any alpha of .10 or higher.

10.38 $H_0: \mu_x - \mu_y = 0; H_1: \mu_x - \mu_y < 0;$ reject Ho if $t_{(44,.05)} < -1.684$

$$s_p^2 = \frac{(n_x - 1)s_x^2 + (n_y - 1)s_y^2}{n_x + n_y - 2} = \frac{22(.055)^2 + 22(.058)^2}{23 + 23 - 2} = .00319$$

$$t = \frac{\overline{X} - \overline{Y} - D_0}{\sqrt{\frac{s_p^2}{n_x} + \frac{s_p^2}{n_y}}} = \frac{.058 - .146}{\sqrt{\frac{.00319}{23} + \frac{.00319}{23}}} = -5.284.$$

Reject H_0 at any common level of alpha

10.40 $H_0: P_x - P_y = 0; H_1: P_x - P_y \neq 0;$

$$\hat{p}_o = \frac{47 + 40}{69 + 69} = .630435, \quad z = \frac{.6812 - .5797}{\sqrt{(.630435)(.369565)(\frac{1}{69} + \frac{1}{69})}} = 1.235,$$

p-value = 2[1-F_Z(1.24)] = 2[1-.8925] = .1075.
Reject H_0 at levels of alpha in excess of 10.75%

10.42 $H_0: \sigma_x = \sigma_y; H_1: \sigma_x \neq \sigma_y;\ s_x = 2.64665, \quad s_y = 1.63561,$

$$F = \frac{s_x^2}{s_y^2} = (2.647)^2/(1.63656)^2 = 2.618.$$

Do not reject H_0 at the 5% level, 2.618 < 5.05 ≈ $F_{(5,5,.05)}$

10.44 a. $H_0: \mu_x - \mu_y = 0; H_1: \mu_x - \mu_y > 0;$

$df = n_1 + n_2 - 2 = 27 + 27 - 2 = 52;\ t_{52,.05} = 1.675$

$$s^2{}_p = \frac{(n_1-1)s_1^{\ 2} + (n_2-1)s_2^{\ 2}}{n_1+n_2-2} = \frac{(27-1)100+(27-1)150}{52} = 125$$

$$t_{calc} = \frac{\bar{x}_2 - \bar{x}_1}{\sqrt{\frac{s^2{}_p}{n_1} + \frac{s^2{}_p}{n_2}}} = \frac{64-60}{\sqrt{\frac{125}{60}+\frac{125}{64}}} = 1.99$$

At the .05 level of significance, reject Ho and accept the alternative that the mean output per hectare is significantly greater with the new procedure.

b. 95% acceptance interval:

$$F_{26,26,.025} = 2.20,\ P(\frac{1}{2.20} \le \frac{s_2^{\ 2}}{s_1^{\ 2}} \le 2.20) = .95,\ F_{calc} = \frac{150}{100} = 1.50$$, because F calc is within the acceptance interval, there is not sufficient evidence against the null hypothesis that the sample variances are not significantly different from each other.

10.46 Assume that the population of matched differences are normally distributed

$H_0: \mu_x - \mu_y = 0; H_1: \mu_x - \mu_y \ne 0;$ reject H_0 if $|t_{(9,.05)}| > 1.833$

$\bar{x}$ of the matched differences = 1.13, s of the matched differences = 1.612

$$t = \frac{1.13-0}{1.612/\sqrt{10}} = 2.22, \text{ p-value } = .054.$$

Reject H_0 at the 10%, but not the 5% level

Paired T-Test and CI: VARIETY A, VARIETY B

```
Paired T for VARIETY A - VARIETY B
             N      Mean     StDev  SE Mean
VARIETY A   10   11.9300    2.9265   0.9254
VARIETY B   10   10.8000    2.5237   0.7981
Difference  10   1.13000   1.61180  0.50969
95% CI for mean difference: (-0.02301, 2.28301)
T-Test of mean difference = 0 (vs not = 0): T-Value = 2.22
P-Value = 0.054
```

10.48 a. $H_0: \mu_x - \mu_y = 0; H_1: \mu_x - \mu_y > 0;$

Results for: Ole.MTW

Two-Sample T-Test and CI: Olesales, Carlsale

```
Two-sample T for Olesales vs Carlsale
              N      Mean     StDev   SE Mean
Olesales    156      3791      5364       429
Carlsale    156      2412      4249       340

Difference = mu Olesales - mu Carlsale
Estimate for difference:   1379
95% lower bound for difference: 475
T-Test of difference = 0 (vs >): T-Value = 2.52   P-Value = 0.006
DF = 310
Both use Pooled StDev =   4839
```

Reject H_0 at the .01 level of significance.

b. $H_0: \mu_x - \mu_y = 0; H_1: \mu_x - \mu_y \neq 0;$

Two-Sample T-Test and CI: Oleprice, Carlpric

```
Two-sample T for Oleprice vs Carlpric
N          Mean      StDev    SE Mean
Oleprice   156      0.819      0.139      0.011
Carlpric   156      0.819      0.120     0.0096

Difference = mu Oleprice - mu Carlpric
Estimate for difference:   -0.0007
95% CI for difference: (-0.0297, 0.0283)
T-Test of difference = 0 (vs not =): T-Value=-0.05 P-Value=0.962  DF=310
Both use Pooled StDev = 0.130
```

Do not reject H_0 at any common level of significance. Note that the 95% confidence interval contains 0, therefore, no evidence of a difference.

10.50 $H_0: P_x - P_y = 0; H_1: P_x - P_y \neq 0;$ reject H_0 if $z < -z_{\alpha/2} = -1.96$ or $z > z_{\alpha/2} = 1.96$

Let $n_x = 270$ and $n_y = 203$. Then, $\hat{p}_x = 56/270 = 0.2074$ and $py = 52/203 = 0.2562$.

$$\hat{p}_0 = \frac{(270)(0.2074) + (203)(0.2562)}{270 + 203} = 0.2283$$

$$z = \frac{0.2074 - 0.2562}{\sqrt{\frac{(0.2283)(1-0.2283)}{270} + \frac{(0.2283)(1-0.2283)}{203}}} = -1.25$$

Do not reject H_0 at the 5% level. Conclude that there is a not a difference in the proportion of humorous ads in British versus American trade magazines.

Chapter 11:

Simple Regression

11.2 a. Prepare a scatter plot.

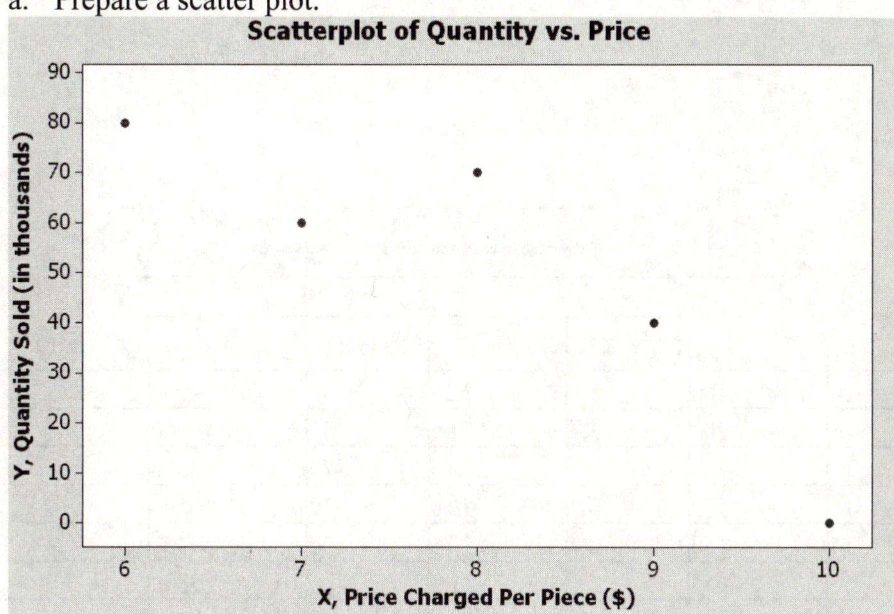

b. Compute the covariance = -45

x_i	y_i	$(x_i - \bar{x})$	$(x_i - \bar{x})^2$	$(y_i - \bar{y})$	$(y_i - \bar{y})^2$	$(x_i - \bar{x})\ (y_i - \bar{y})$
6	80	-2	4	30	900	-60
7	60	-1	1	10	100	-10
8	70	0	0	20	400	0
9	40	1	1	-10	100	-10
10	0	2	4	-50	2500	-100
40	250	0	10	0	4000	-180
$\bar{x}$ = 8.00	$\bar{y}$ = 50.00		s_x^2 = 2.5		s_y^2 =1000	Cov(x,y) = -45
			s_x = 1.5811		s_y =31.623	

c. Compute and interpret $b_1 = \dfrac{Cov(x,y)}{s_x^2} = \dfrac{-45}{2.5} = -18.0$. For a one dollar increase in the price per piece of plywood, the quantity sold of plywood is estimated to decrease by 18 thousand pieces.

d. Compute $b_0 = \bar{y} - b_1\bar{x}$ = 50.0 – (-18)(8.0) = 194.00

e. What quantity of plywood is expected to be sold if the price were $7 per piece?

$\hat{y} = b_0 + b_1 x = 194.00 + -18.0(7) = 68$ thousand pieces sold

11.4 a. Scatter plot – Advertising expenditures (thousands of $s) vs. Monthly Sales (thousands of units)

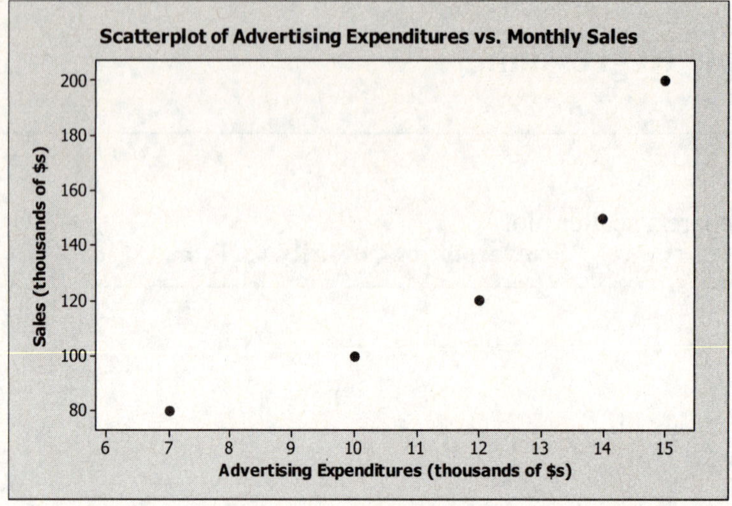

x_i	y_i	$(x_i-\bar{x})$	$(x_i-\bar{x})^2$	$(y_i-\bar{y})$	$(y_i-\bar{y})^2$	$(x_i-\bar{x})(y_i-\bar{y})$
10	100	-1.6	2.56	-30	900	48
15	200	3.4	11.56	70	4900	238
7	80	-4.6	21.16	-50	2500	230
12	120	0.4	0.16	-10	100	-4
14	150	2.4	5.76	20	400	48
58	650		41.2		8800	560
$\bar{x}$ = 11.60	$\bar{y}$ = 130.00		s_x^2 = 10.3		s_y^2 =2200	Cov(x,y) = 140
			s_x = 3.2094		s_y = 46.9042	
				b1 =	13.5922	
				b0 =	-27.67	

$$\text{Covariance} = Cov(x,y) = \frac{\sum (x_i-\bar{x})(y_i-\bar{y})}{n-1} = 560 / 4 = 140$$

$$\text{Correlation} = \frac{Cov(x,y)}{s_x s_y} = \frac{140}{(3.2094)(46.9042)} = .93002$$

b. It appears that advertising has a positive effect on sales. Note that correlation does not imply causation. Other factors could have been changing at the same time that advertising changed. For example, prices of competitive goods could have been changing or tastes and preferences of consumers, or the number of buyers in the market.

c. Compute the regression $b_1 = \frac{Cov(x,y)}{s_x^2}$ = 140 / 10.3 = 13.5922 and coefficients $b_0 = \bar{y} - b_1\bar{x}$ = 130.0 – (13.5922)(11.6) = -27.669

11.6 a. Scatter plot

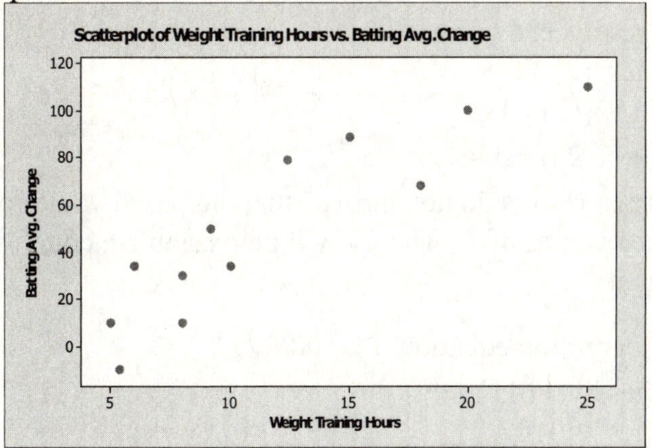

The data shows a positive relationship between the average number of hours spent in the weight training program versus the increment to each player's batting average. It appears that the weight training program has been effective, although correlation does not necessarily imply causation.

b. Estimate the regression equation.

x_i	y_i	$(x_i-\bar{x})$	$(x_i-\bar{x})^2$	$(y_i-\bar{y})$	$(y_i-\bar{y})^2$	$(x_i-\bar{x})(y_i-\bar{y})$
8	10	-3.83333	14.69444	-40.3333	1626.778	154.61111
20	100	8.166667	66.69444	49.66667	2466.778	405.61111
5.4	-10	-6.43333	41.38778	-60.3333	3640.111	388.14444
12.4	79	0.566667	0.321111	28.66667	821.7778	16.244444
9.2	50	-2.63333	6.934444	-0.33333	0.111111	0.8777778
15	89	3.166667	10.02778	38.66667	1495.111	122.44444
6	34	-5.83333	34.02778	-16.3333	266.7778	95.277778
8	30	-3.83333	14.69444	-20.3333	413.4444	77.944444
18	68	6.166667	38.02778	17.66667	312.1111	108.94444
25	110	13.16667	173.3611	59.66667	3560.111	785.61111
10	34	-1.83333	3.361111	-16.3333	266.7778	29.944444
5	10	-6.83333	46.69444	-40.3333	1626.778	275.61111
142	604		450.2267		16496.67	2461.2667

$\bar{x} = 11.83$ $\bar{y} = 50.33$ $s_x^2 = 40.9297$ $s_y^2 =$ 1499.697 Cov(x,y) = 223.75152

6.397632 38.72592 $s_x = 6.397632$ $s_y = 38.72592$

$$\text{Covariance} = Cov(x,y) = \frac{\sum(x_i-\bar{x})(y_i-\bar{y})}{n-1} = 2461.2667 / 11 = 223.75152$$

$$\text{Correlation} = \frac{Cov(x,y)}{s_x s_y} = \frac{223.75152}{(6.397632)(38.72592)} = .90312$$

$$\text{Regression coefficients } b_1 = \frac{Cov(x,y)}{s_x^2} = 223.75152 / 40.9297 = 5.4667$$

and $b_0 = \bar{y} - b_1\bar{x} = 50.33 - (5.4667)(11.83) = -14.3411$

Regression equation = $\hat{Y}$ = -14.4 + 5.47 X. Each additional hour of the weight training program yields an expected improvement in batting average of 5.47 points.

11.8 Given the regression equation $\hat{y} = -50 + 12X$

a. Y changes by +36
b. Y changes by -48
c. $\hat{y} = 50 + 12(12) = 194$
d. $\hat{y} = 50 + 12(23) = 326$
e. Regression results do not "prove" that increased values of X "causes" increased values of Y. Theory will help establish conclusions of causation.

11.10 Given the regression equation $\hat{y} = 100 + 21X$

a. Y changes by +105
b. Y changes by -147
c. $\hat{y} = 100 + 21(14) = 394$
d. $\hat{y} = 100 + 21(27) = 667$
e. Regression results do not "prove" that increased values of X "causes" increased values of Y. Theory will help establish conclusions of causation.

11.12 There are insufficient data points to create an accurate regression equation. The estimate is significantly outside previous data values so any estimate would be faulty and not reliable.

11.14 A population regression equation consists of the true regression coefficients β_i and the true model error ε_i. By contrast, the estimated regression model consists of the estimated regression coefficients b_i and the residual term e_i. The population regression equation is a model that purports to measure the actual value of Y, while the sample regression equation is an estimate of the predicted value of the dependent variable Y.

11.16 The constant represents an adjustment for the estimated model and not the number sold when the price is zero.

11.18 Compute the coefficients for a least squares regression equation

$$b_1 = r_{xy}\frac{s_y}{s_x} \quad b_0 = \overline{y} - b_1\overline{x} \quad \hat{y}_i = b_0 + b_1 x_i$$

a. $b_1 = .6\frac{75}{25} = 1.80 \quad b_0 = 100 - 1.80(50) = 10 \quad \hat{y}_i = 10 + 1.80x_i$

b. $b_1 = .7\frac{65}{35} = 1.30 \quad b_0 = 210 - 1.30(60) = 132 \quad \hat{y}_i = 132 + 1.30x_i$

c. $b_1 = .75\frac{78}{60} = .975 \quad b_0 = 100 - .975(20) = 80.5 \quad \hat{y}_i = 80.5 + .975x_i$

d. $b_1 = .4\frac{75}{100} = .30 \quad b_0 = 50 - .30(10) = 47 \quad \hat{y}_i = 47 + .30x_i$

e. $b_1 = .6\frac{70}{80} = .525 \quad b_0 = 200 - .525(90) = 152.75 \quad \hat{y}_i = 152.75 + .525x_i$

11.20 a. $n = 20, \overline{X} = 25.4/20 = 1.27, \overline{Y} = 22.6/20 = 1.13$

$$b_1 = \frac{150.5 - (20)(1.13)(1.27)}{145.7 - (20)(1.27)^2} = 1.0737, \; b_0 = 1.13 - 1.0737(1.27) = -.2336$$

b. For a one unit increase in the rate of return of the S&P 500 index, we estimate that the rate of return of the corporation's stock will increase by 1.07%

c. When the percentage rate of return of the S&P 500 index is zero, we estimate that the corporation's rate of return will be -.2336%

11.22 a. $b_1 = 180/350 = .5143$, $b_0 = 16 - .5143(25.5) = 2.8854$ $\hat{y} = 2.8854 + .5143x$

b. For a one unit increase in the average cost of a meal, we would estimate that the number of bottles sold would increase by .5148%

c. Yes. 2.8854 bottles are estimated to be sold, regardless of the price paid for a meal.

11.24 a. $\hat{y} = 1.89 + 0.0896x$

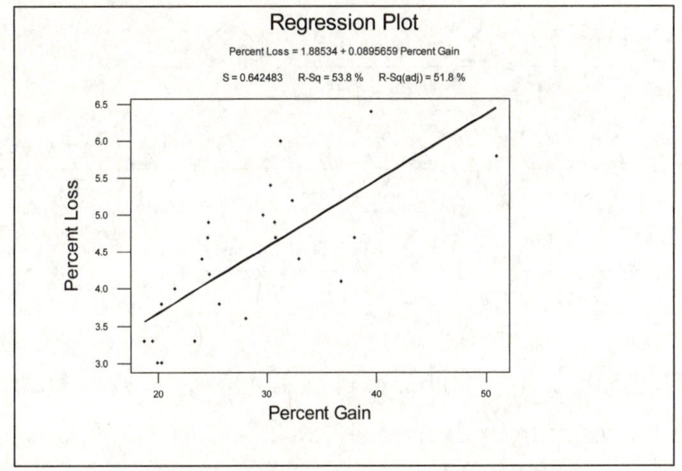

b. 0.0896%. For a one percent pre-November 13 gain, we would estimate that there would be a loss of .0896% on November 13.

11.26 Compute SSR, SSE, s^2_e and the coefficient of determination.

Note that $R^2 = r^2_{xy}$, $R^2 = \frac{SSR}{SST} = 1 - \frac{SSE}{SST}$, and $\hat{\sigma}^2 = s_e^{\,2} = \frac{SSE}{n-2}$

a. $.5 = \frac{SSR}{100{,}000}$ Solve for SSR. SSR = 50,000. Therefore, given that SST = SSR + SSE, 100,000 = 50,000 + SSE. Solve for SSE. SSE = 50,000.

$$\hat{\sigma}^2 = s_e^{\,2} = \frac{50{,}000}{52-2} = 1{,}000$$

b. $.7 = \frac{SSR}{90{,}000}$ Solve for SSR. SSR = 63,000. Therefore, given that SST = SSR + SSE, 90,000 = 63,000 + SSE. Solve for SSE.

SSE = 27,000. $\hat{\sigma}^2 = s_e^{\,2} = \frac{27{,}000}{52-2} = 540$

c. $.8 = \frac{SSR}{240}$ Solve for SSR. SSR = 192. Therefore, given that SST = SSR + SSE, 240 = 192 + SSE. Solve for SSE. SSE = 48. $\hat{\sigma}^2 = s_e^{\ 2} = \frac{48}{52-2} = .96$

d. $.3 = \frac{SSR}{200,000}$ Solve for SSR. SSR = 60,000. Therefore, given that SST = SSR + SSE, 200,000 = 60,000 + SSE. Solve for SSE. SSE = 140,000. $\hat{\sigma}^2 = s_e^{\ 2} = \frac{140,000}{74-2} = 1,944.444$

e. $.9 = \frac{SSR}{60,000}$ Solve for SSR. SSR = 54,000. Therefore, given that SST = SSR + SSE, 60,000 = 54,000 + SSE. Solve for SSE. SSE = 6,000. $\hat{\sigma}^2 = s_e^{\ 2} = \frac{6,000}{40-2} = 157.8947$

11.28 a. $$R^2 = \frac{\sum(\hat{y}_i - \bar{y})^2}{\sum(y_i - \bar{y})^2} = \frac{\sum[(b_i(x_i - \bar{x})]^2}{\sum(y_i - \bar{y})^2} = b_i^{\ 2}\frac{\sum(x_i - \bar{x})^2}{\sum(y_i - \bar{y})^2}$$

b. $$R^2 = b_i^{\ 2}\frac{\sum(x_i - \bar{x})^2}{\sum(y_i - \bar{y})^2} = b\frac{\sum(x_i - \bar{x})(y_i - \bar{y})}{\sum(y_i - \bar{y})^2} = \frac{[\sum(x_i - \bar{x})(y_i - \bar{y})]^2}{\sum(x_i - x)^2\sum(y_i - y)^2} = r^2$$

c. $$b_1 b_1^{\ *} = \frac{\sum(x_i - \bar{x})(y_i - \bar{y})}{\sum(x_i - \bar{x})^2}\frac{\sum(x_i - \bar{x})(y_i - \bar{y})}{\sum(y_i - \bar{y})^2} = r^2$$

11.30 $n = 13, \bar{x} = .5538, \sum x^2 = 80.06, \bar{y} = 11.8154, \sum y^2 = 3718.76, b = -2.0341$

Based on the result from Exercise 11.28 a:

$$R^2 = (2.0341)^2 \frac{80.06 - 13(.5538)^2}{3718.76 - 13(11.8154)^2} = .1653$$

Using the computer, we find that r = -.4066, $r^2 = .1653 = R^2$

Correlations: Dow5day, Dow1Yr

```
Pearson correlation of Dow5day and Dow1Yr = -0.407
P-Value = 0.168
```

Or, from the Minitab output:

Regression Analysis: Dow1Yr versus Dow5day

```
The regression equation is
Dow1Yr = 12.9 - 2.03 Dow5day
Predictor       Coef    SE Coef        T      P
Constant      12.942      3.420     3.78  0.003
Dow5day       -2.034      1.378    -1.48  0.168

S = 12.02       R-Sq = 16.5%     R-Sq(adj) = 8.9%

Analysis of Variance
Source          DF        SS        MS       F       P
Regression       1     314.8     314.8    2.18   0.168
Residual Error  11    1589.2     144.5
Total           12    1903.9
```

Squaring the simple correlation coefficient of r = -.4066 yields the coefficient determination for the corresponding bivariate regression $r^2 = .1653$

11.32 a. The Minitab output below shows the predicted value (Fit) and the residuals:

Regression Analysis: Change in Mean absence illness versus Change in Absentee rate

```
The regression equation is
Change in Mean absence illness = 0.0449 - 0.224 Change in Absentee Rate
Predictor                      Coef   SE Coef       T      P
Constant                    0.04485   0.06347    0.71  0.498
Change in Absentee Rate    -0.22426   0.05506   -4.07  0.003

S = 0.207325   R-Sq = 64.8%   R-Sq(adj) = 60.9%
Analysis of Variance
Source          DF        SS        MS      F      P
Regression       1   0.71315   0.71315  16.59  0.003
Residual Error   9   0.38685   0.04298
Total           10   1.10000

Unusual Observations
       Change    Change
           in   in Mean
     Absentee   absence
Obs      Rate   illness      Fit   SE Fit  Residual  St Resid
  3      1.40    0.2000  -0.2691   0.0910    0.4691     2.52R
  8      2.90   -0.8000  -0.6055   0.1613   -0.1945    -1.49 X

R denotes an observation with a large standardized residual.
X denotes an observation whose X value gives it large influence.
```

b. $SST = \sum y^2 - n\bar{y}^2 = 1.1 - 25(0.0)^2 = 1.1$

$SSR = \sum(\hat{y}_i - \bar{y})^2 = .713$

$SSE = \sum e_i^2 = .387$

$SST = 1.1 = .713 + .387 = SSR + SSE$

c. $R^2 = SSR/SST = .713/1.1 = .648$, 64.8% of the variation in the dependent variable mean employee absence rate due to own illness can be explained by the variation in the change in absentee rate.

11.34 $R^2 = r^2 = (.11)^2 = .0121$. 1.21% of the variation in the dependent variable annual raises can be explained by the variation in teaching evaluations.

11.36 For a simple regression problem, test the hypothesis $H_o: \beta_1 = 0 \text{ vs } H_1: \beta_1 \neq 0$

Given that $R^2 = r^2_{xy}$, $R^2 = \frac{SSR}{SST} = 1 - \frac{SSE}{SST}$, and $\hat{\sigma}^2 = s_e^2 = \frac{SSE}{n-2}$

a. n=35, SST = 100,000, r = .46

$R^2 = (.46)^2 = .2116$. $.2116 = \frac{SSR}{100,000}$ SSR = 21,160.

Therefore, given that SST = SSR + SSE, 100,000 = 21,160 + SSE.

SSE = 78,840. $\hat{\sigma}^2 = s_e^2 = \frac{78,840}{35-2} = 2,389.091$

$F = \frac{MSR}{MSE} = \frac{SSR}{s^2_e} = \frac{21,160}{2389.091} = 8.857$. $F_{\alpha,1,n-2} = t^2_{\alpha/2,n-2} = 2.042^2 = 4.170$

Therefore, at the .05 level, Reject H_0

b. $R^2 = (.65)^2 = .4225$. $.4225 = \dfrac{SSR}{123,000}$ SSR = 51,967.5.

Given that SST = SSR + SSE, 123,000 = 51,967.5 + SSE.

SSE = 71,032.5. $\hat{\sigma}^2 = s_e^{\ 2} = \dfrac{71,032.5}{61-2} = 1,203.941$

$$F = \frac{MSR}{MSE} = \frac{SSR}{s^2_{\ e}} = \frac{51,967.5}{1203.941} = 43.165.\ \ F_{\alpha,1,n-2} = t^2_{\ \alpha/2,n-2} = 2.000^2 = 4.00$$

Therefore, at the .05 level, Reject H_0

c. $R^2 = (.69)^2 = .4761$. $.4761 = \dfrac{SSR}{128,000}$ SSR = 60,940.8.

Given that SST = SSR + SSE, 128,000 = 60,940.8 + SSE. SSE = 67,059.2. $\hat{\sigma}^2 = s_e^{\ 2} = \dfrac{67,059.2}{25-2} = 2,915.617$

$$F = \frac{MSR}{MSE} = \frac{SSR}{s^2_{\ e}} = \frac{60,940.8}{2915.617} = 20.902.\ \ F_{\alpha,1,n-2} = t^2_{\ \alpha/2,n-2} = 2.069^2 = 4.281$$

Therefore, at the .05 level, Reject H_0

11.38 a. $n = 8, \bar{X} = 52/8 = 6.5, \sum x^2 = 494$,

$\bar{Y} = 54.4/8 = 6.8, \sum y^2 = 437.36, \sum xy = 437.7$,

$$b_1 = \frac{437.7 - 8(6.5)(6.8)}{494 - 8(6.5)^2} = .5391$$

$b_0 = 6.8 - .5391(6.5) = 3.2958$

b. $\sum e_i^{\ 2} = [437.36 - 8(6.8)^2] - (.5391)^2[494 - 8(6.5)^2] = 22.1019$

$s_e^{\ 2} = 22.1019/6 = 3.6836$, $s_b^{\ 2} = \dfrac{3.6836}{494 - 8(6.5)^2} = .0236$, $t_{6,.05} = 1.943$,

Therefore, the 90% confidence interval is: $.5391 \pm 1.943\sqrt{.0236}$, .2406 up to .8376

Regression Analysis: SalesChg_Ex11.38 versus AdvertChg_Ex11.38

```
The regression equation is
SalesChg_Ex11.38 = 3.30 + 0.539 AdvertChg_Ex11.38
Predictor              Coef  SE Coef     T      P
Constant              3.296    1.208  2.73  0.034
AdvertChg_Ex11.38    0.5391   0.1537  3.51  0.013

S = 1.91927   R-Sq = 67.2%   R-Sq(adj) = 61.8%
Analysis of Variance
Source           DF      SS      MS      F      P
Regression        1  45.339  45.339  12.31  0.013
Residual Error    6  22.101   3.684
Total             7  67.440
```

11.40 **Regression Analysis: Dow1Yr versus Dow5day**

```
The regression equation is
Dow1Yr = 12.9 - 2.03 Dow5day
Predictor     Coef  SE Coef      T      P
Constant    12.942    3.420   3.78  0.003
Dow5day     -2.034    1.378  -1.48  0.168

S = 12.0195   R-Sq = 16.5%   R-Sq(adj) = 8.9%
Analysis of Variance
Source          DF      SS     MS     F      P
Regression       1   314.8  314.8  2.18  0.168
Residual Error  11  1589.2  144.5
Total           12  1903.9
```

a. $\sum e_i^2 = SSE = 1589.2$

$s_e^2 = SSE/(n-2) = 1589.2/11 = 144.4686$

b. $s_b^2 = (1.378)^2 = 1.899$

c. $t_{11,.025} = 2.201$.

Therefore, the 95% confidence interval is: -2.0341 ± 2.201(1.378), -5.0673 up to .9991

d. $H_o : \beta = 0; H_1 : \beta \neq 0;$

$$t = \frac{-2.0341}{1.378} = -1.48$$

Therefore, do not reject H_0 at the 10% level since t = -1.48 > -1.796 = $-t_{11,.05}$

11.42 Given a simple regression: $\hat{y}_{n+1} = 12 + 5(13) = 77$

95% Prediction Interval: $\hat{y}_{n+1} \pm t_{n-2,\alpha/2}\sqrt{\left[1 + \frac{1}{n} + \frac{(x_{n+1}-\bar{x})^2}{\sum(x_i-\bar{x})^2}\right]}(s_e)$

$$77 \pm 2.042\sqrt{\left[1 + \frac{1}{32} + \frac{(13-8)^2}{500}\right]}(9.67) = 77 \pm 20.533,\ (56.467,\ 97.533)$$

95% Confidence Interval: $\hat{y}_{n+1} \pm t_{n-2,\alpha/2}\sqrt{\left[\frac{1}{n} + \frac{(x_{n+1}-\bar{x})^2}{\sum(x_i-\bar{x})^2}\right]}(s_e)$

$$77 \pm 2.042\sqrt{\left[\frac{1}{32} + \frac{(13-8)^2}{500}\right]}(9.67) = 77 \pm 5.629,\ (71.371,\ 82.629)$$

11.44 Given a simple regression: $\hat{y}_{n+1} = 22 + 8(17) = 158$

95% Prediction Interval: $\hat{y}_{n+1} \pm t_{n-2,\alpha/2}\sqrt{\left[1 + \frac{1}{n} + \frac{(x_{n+1}-\bar{x})^2}{\sum(x_i-\bar{x})^2}\right]}(s_e)$

$$158 \pm 2.086\sqrt{\left[1 + \frac{1}{22} + \frac{(17-11)^2}{400}\right]}(3.45) = 158 \pm 7.669,\ (150.331,\ 165.669)$$

95% Confidence Interval: $\hat{y}_{n+1} \pm t_{n-2,\alpha/2}\sqrt{\left[\frac{1}{n} + \frac{(x_{n+1}-\bar{x})^2}{\sum(x_i-\bar{x})^2}\right]}(s_e)$

$$158 \pm 2.086\sqrt{\left[\frac{1}{22}+\frac{(17-11)^2}{400}\right](3.45)} = 158 \pm 2.649,\ (155.351,\ 160.649)$$

11.46 $s_e^{\,2} = 80.6/23 = 3.5043$, $s_b^{\,2} = \dfrac{3.5043}{130} = .027$

a. $H_o: \beta = 0; H_1: \beta < 0;,\ t = \dfrac{-1.2}{\sqrt{.027}} = -7.303$

Therefore, reject H_0 at the 1% level since t = -7.303 > -2.807 = -$t_{23,.005}$

b. $y_{n+1} = 12.6 - 1.2(4) = 7.8$, $7.8 \pm 1.714\sqrt{\left[1+\frac{1}{25}+\frac{(4-6)^2}{130}\right](1.872)}$

7.8 ± 3.3203, (4.4798, 11.1203)

11.48 a. $R^2 = 1 - \dfrac{SSE}{SST}, SSE = SST(1-R^2), R^2 = .7662$

$SST = \sum y^2 - n\bar{y}^2 = 196.2 - 20(1.13)^2 = 170.662$,

$SSE = 170.662(1-.7662) = 39.9008$

$s^2{}_e = 39.9008/18 = 2.2167$, $s^2{}_b = \dfrac{2.2167}{145.7 - 20(1.27)^2} = .0195$

$H_o: \beta = 0, H_1: \beta > 0,\ t = \dfrac{1.0737}{\sqrt{.0195}} = 7.689$,

Therefore, reject H_0 at the 1% level since t = 7.689 > 2.878 = $t_{18,.005}$

b. $H_o: \beta = 1, H_1: \beta \neq 1,\ t = \dfrac{1.0737-1}{\sqrt{.0195}} = .5278$

Therefore, do not reject H_0 at the 20% level since t = .5278 < 1.33 = $t_{18,.10}$

11.50 $s^2{}_b = .000299, b = .0896$

$H_o: \beta = 0, H_1: \beta \neq 0,\ t = \dfrac{.0896}{\sqrt{.000299}} = 5.1817$,

Therefore, reject H_0 at the 1% level since t = 5.1817 > 2.807 = $t_{23,.005}$

11.52 $\sum (y_i - \bar{y})^2 = 250(16) = 4000$

$\sum (x_i - \bar{x})^2 = 350(16) = 5600$

$SSE = 4000 - (.5143)^2(5600) = 2518.7749$, $s^2{}_e = 2518.7749/15 = 167.9183$

$s^2{}_b = 167.9183/5600 = .03$, $H_o: \beta = 0, H_1: \beta \neq 0$, $t = \dfrac{.5143}{\sqrt{.03}} = 2.969$

Therefore, reject H_0 at the 1% level since t = 2.969 > 2.947 = $t_{15,.005}$

11.54 $H_o: \beta = 0, H_1: \beta < 0,\ t = \dfrac{-5.903}{\sqrt{.3316}} = -10.251$

Therefore, reject H_0 at the .5% level since t = -10.251 < -3.707 = $t_{6,.005}$

11.56 $\hat{Y}_{n+1} = 12.942 - 2.034(1) = 10.9079$

b. $t_{11,.05} = 1.796$

The 90% prediction interval for prediction of the actual value:

$$10.9079 \pm 1.796\sqrt{[1 + \frac{1}{13} + \frac{(1-.5538)^2}{80.06 - 13(.5538)^2}](12.0195)}$$

10.9079 ± 22.4291, -11.5212 up to 33.337

The 90% confidence interval for prediction of the expected value:

$$10.9079 \pm 1.796\sqrt{[\frac{1}{13} + \frac{(1-.5538)^2}{80.06 - 13(.5538)^2}](12.0195)}$$

10.9070 ± 6.0882, 4.8197 up to 16.9961

The prediction interval is the estimate of the actual value that results for a single observation of the independent variable. The confidence interval estimates the conditional mean, that is, the average value of the dependent variable when the independent variable is fixed at a specific level.

11.58 $\hat{Y}_{n+1} = -.2336 + 1.0737(1) = .8401$, $t_{18,.05} = 1.734$

The 90% confidence interval for prediction of the expected value:

$.8401 \pm 1.734\sqrt{[\frac{1}{20} + \frac{(1-1.27)^2}{145.7 - 20(1.27)^2}](1.4889)}$, .8401 ±.581, .2591 up to .14211

95% confidence interval for the prediction of the expected value:

$.8401 \pm 2.101\sqrt{[\frac{1}{20} + \frac{(1-1.27)^2}{145.7 - 20(1.27)^2}](1.4889)}$, .8401 ±.7039, .1362 up to 1.544

11.60 a. Compute the sample correlation

x	y	$(x_i - \bar{x})$	$(x_i - \bar{x})^2$	$(y_i - \bar{y})$	$(y_i - \bar{y})^2$	$(x_i - \bar{x})(y_i - \bar{y})$
2	5	-1.8	3.24	-2.4	5.76	4.32
5	8	1.2	1.44	0.6	0.36	0.72
3	7	-0.8	0.64	-0.4	0.16	0.32
1	2	-2.8	7.84	-5.4	29.16	15.12
8	15	4.2	17.64	7.6	57.76	31.92
19	**37**		**30.8**		**93.2**	**52.4**

$\bar{x}$ = 19/5 = 3.8, $\bar{y}$ = 37/5 = 7.4, $s_x = \sqrt{\frac{\sum(x_i - \bar{x})^2}{n-1}} = \sqrt{\frac{30.8}{4}} = 2.7749$,

$s_y = \sqrt{\frac{\sum(y_i - \bar{y})^2}{n-1}} = \sqrt{\frac{93.2}{4}} = 4.827$, $s_{xy} = \frac{\sum(x_i - \bar{x})(y_i - \bar{y})}{n-1} = 52.4/4 = 13.1$

$r = \frac{s_{xy}}{s_x s_y}$ = 13.1/(2.7749)(4.827) = .97802

b. Compute the sample correlation

x	y	$(x_i-\bar{x})$	$(x_i-\bar{x})^2$	$(y_i-\bar{y})$	$(y_i-\bar{y})^2$	$(x_i-\bar{x})(y_i-\bar{y})$
7	5	-1.8	3.24	-2.4	5.76	4.32
10	8	1.2	1.44	0.6	0.36	0.72
8	7	-0.8	0.64	-0.4	0.16	0.32
6	2	-2.8	7.84	-5.4	29.16	15.12
13	15	4.2	17.64	7.6	57.76	31.92
44	**37**	**0**	**30.8**	**0**	**93.2**	**52.4**

$$\bar{x} = 44/5 = 8.8,\ \bar{y} = 37/5 = 7.4,\ s_x = \sqrt{\frac{\sum(x_i-\bar{x})^2}{n-1}} = \sqrt{\frac{30.8}{4}} = 2.7749,$$

$$s_y = \sqrt{\frac{\sum(y_i-\bar{y})^2}{n-1}} = \sqrt{\frac{93.2}{4}} = 4.827,\ s_{xy} = \frac{\sum(x_i-\bar{x})(y_i-\bar{y})}{n-1} = 52.4/4 = 13.1$$

$$r = \frac{s_{xy}}{s_x s_y} = 13.1/(2.7749)(4.827) = .97802$$

c. Compute the sample correlation

x	y	$(x_i-\bar{x})$	$(x_i-\bar{x})^2$	$(y_i-\bar{y})$	$(y_i-\bar{y})^2$	$(x_i-\bar{x})(y_i-\bar{y})$
12	4	-3.6	12.96	-1.8	3.24	6.48
15	6	-0.6	0.36	0.2	0.04	-0.12
16	5	0.4	0.16	-0.8	0.64	-0.32
21	8	5.4	29.16	2.2	4.84	11.88
14	6	-1.6	2.56	0.2	0.04	-0.32
78	**29**		**45.2**		**8.8**	**17.6**

$$\bar{x} = 78/5 = 15.6,\ \bar{y} = 29/5 = 5.8,\ s_x = \sqrt{\frac{\sum(x_i-\bar{x})^2}{n-1}} = \sqrt{\frac{45.2}{4}} = 3.36155,$$

$$s_y = \sqrt{\frac{\sum(y_i-\bar{y})^2}{n-1}} = \sqrt{\frac{8.8}{4}} = 1.48324,\ s_{xy} = \frac{\sum(x_i-\bar{x})(y_i-\bar{y})}{n-1} = 17.6/4 = 4.4$$

$$r = \frac{s_{xy}}{s_x s_y} = 4.4/(3.36155)(1.48324) = .88247$$

d. Compute the correlation coefficient

x	y	$(x_i - \bar{x})$	$(x_i - \bar{x})^2$	$(y_i - \bar{y})$	$(y_i - \bar{y})^2$	$(x_i - \bar{x})(y_i - \bar{y})$
2	8	-1.8	3.24	-5	25	9
5	12	1.2	1.44	-1	1	-1.2
3	14	-0.8	0.64	1	1	-0.8
1	9	-2.8	7.84	-4	16	11.2
8	22	4.2	17.64	9	81	37.8
19	**65**	**0**	**30.8**	**0**	**124**	**56**

$\bar{x} = 19/5 = 3.8,\ \bar{y} = 65/5 = 13,\ s_x = \sqrt{\dfrac{\sum (x_i - \bar{x})^2}{n-1}} = \sqrt{\dfrac{30.8}{4}} = 2.77488,$

$s_y = \sqrt{\dfrac{\sum (y_i - \bar{y})^2}{n-1}} = \sqrt{\dfrac{124}{4}} = 5.56776,\ s_{xy} = \dfrac{\sum (x_i - \bar{x})(y_i - \bar{y})}{n-1} = 56/4 = 14$

$r = \dfrac{s_{xy}}{s_x s_y} = 14/(2.77488)(5.56776) = .90615$

11.62 Let x = Examination and y = Project

x	y	$(x_i - \bar{x})$	$(x_i - \bar{x})^2$	$(y_i - \bar{y})$	$(y_i - \bar{y})^2$	$(x_i - \bar{x})(y_i - \bar{y})$
81	76	2.4	5.76	-0.7	0.49	-1.68
62	71	-16.6	275.56	-5.7	32.49	94.62
74	69	-4.6	21.16	-7.7	59.29	35.42
78	76	-0.6	0.36	-0.7	0.49	0.42
93	87	14.4	207.36	10.3	106.09	148.32
69	62	-9.6	92.16	-14.7	216.09	141.12
72	80	-6.6	43.56	3.3	10.89	-21.78
83	75	4.4	19.36	-1.7	2.89	-7.48
90	92	11.4	129.96	15.3	234.09	174.42
84	79	5.4	29.16	2.3	5.29	12.42
786	**767**		**824.4**		**668.1**	**575.8**

$\bar{x} = 78.6,\ \bar{y} = 76.7,\ s_x = \sqrt{\dfrac{\sum (x_i - \bar{x})^2}{n-1}} = \sqrt{\dfrac{824.4}{9}} = 9.5708,$

$s_y = \sqrt{\dfrac{\sum (y_i - \bar{y})^2}{n-1}} = \sqrt{\dfrac{668.1}{9}} = 8.6159,\ s_{xy} = \dfrac{\sum (x_i - \bar{x})(y_i - \bar{y})}{n-1} = 575.8/9 =$

$63.97778,\ r = \dfrac{s_{xy}}{s_x s_y} = 63.97778/(9.5708)(8.6159) = .7759$

11.64 $H_o: \rho = 0, H_1: \rho > 0, t = \dfrac{r\sqrt{(n-2)}}{\sqrt{(1-r^2)}} = t = \dfrac{.11\sqrt{351}}{\sqrt{1-(.11^2)}} = 2.073$

Therefore, reject H_0 at the 2.5% level since $2.073 > 1.96 = z_{.025} \approx t_{351,.025}$

11.66 a. Using the computer, calculate the sample correlation

Correlations: Dow5day, Dow1Yr

```
Pearson correlation of Dow5day and Dow1Yr = -0.407
P-Value = 0.168
```

b. $H_o: \rho = 0, H_1: \rho \neq 0$, $t = \frac{-.4066\sqrt{11}}{\sqrt{1-(-.4066)^2}} = -1.4761$

$t_{11,.05} = \pm 1.796$, p-value of .168 > alpha of .10. Do not reject H_0 at the 10% level

11.68 Using the computer, calculate the correlation and test against a two-sided alternative

Results for: Advertising Revenue.MTW

Correlations: Cost of advertisement, Revenue from Inquiries

```
Pearson correlation of X and Y = 0.0575
P-Value = 0.827
```

$H_o: \rho = 0, H_1: \rho \neq 0$, $t = \frac{.0575\sqrt{15}}{\sqrt{1-(.0575)^2}} = .2231$,

Do not reject H_0 at the 20% level since $.2231 < 1.341 = t_{15,.10}$

11.70 The Senior Housing return has a beta of 1.369 with a coefficient Student's $t = 3.87$ and an overall R-squared of 20.5%. This means the nondiversifiable risk response for Senior Housing is significantly above the overall market. This firm's return is more responsive to the market.

Senior Housing	mean	0.008144
SUMMARY OUTPUT	return	0.81%

Regression Statistics	
Multiple R	0.45332
R Square	0.205499
Adjusted R Square	0.191801
Standard Error	0.068181
Observations	60

ANOVA

	df	*SS*	*MS*	*F*	*Significance F*
Regression	1	0.069738	0.069738	15.0018	0.000275
Residual	58	0.269623	0.004649		
Total	59	0.339362			

	Coefficients	*Standard Error*	*t Stat*	*P-value*	*Lower 95%*	*Upper 95%*
Intercept	-0.00082	0.009101	-0.08991	0.928668	-0.01904	0.0174
SP 500	1.368854	0.353415	3.873216	0.000275	0.661417	2.076292

11.72 The Seagate return has a beta of 1.810 with a coefficient Student's $t = 2.98$ and an overall R-squared of 13.2%. The nondiversifiable risk response for Seagate is substantially above the overall market. For the 60-month period the average monthly return for Seagate was –0.03%, with the average variance equal to 0.0156.
The Microsoft return has a beta of 0.967 with a coefficient Student's $t = 3.31$ and an overall R-squared of 15.9%. The nondiversifiable risk response for Microsoft is slightly below the overall market. For the 60-month period the average monthly return for Microsoft was 0.00%, with the average variance equal to 0.0037.

The Tata return has a beta of 2.796 with a coefficient Student's $t = 4.14$ and an overall R-squared of 22.8%. The nondiversifiable risk response for Tata is extremely above the overall market. For the 60-month period the average monthly return for Tata was 2.23%, with the average variance equal to 0.0216.
The Tata stock has the most risk and the most return. Microsoft is the least risky but has a low return. If the market is trending upward Tata is the recommended stock. If the market is volatile or trending downward Microsoft is the recommended stock.

Seagate	mean	-0.00026676	-0.03%
	Variance	0.01560793	
Microsoft	mean	4.59393E-07	0.00%
	Variance	0.003712483	
Tata	mean	0.022260121	2.23%
	Variance	0.021645412	

SUMMARY OUTPUT Seagate!

Regression Statistics	
Multiple R	0.363972
R Square	0.132476
Adjusted R Square	0.117518
Standard Error	0.117361
Observations	60

ANOVA

	df	*SS*	*MS*	*F*	*Significance F*
Regression	1	0.121992712	0.121993	8.856925	0.004253
Residual	58	0.798875163	0.013774		
Total	59	0.920867875			

	Coefficients	*Standard Error*	*t Stat*	*P-value*	*Lower 95%*	*Upper 95%*
Intercept	-0.01212	0.015666099	-0.77368	0.44226	-0.04348	0.019238
SP 500	1.810456	0.608340183	2.976059	0.004253	0.592732	3.02818

SUMMARY OUTPUT Microsoft

Regression Statistics	
Multiple R	0.398804
R Square	0.159045
Adjusted R Square	0.144546
Standard Error	0.056355
Observations	60

ANOVA

	df	*SS*	*MS*	*F*	*Significance F*
Regression	1	0.034836621	0.034837	10.96919	0.001598
Residual	58	0.184199855	0.003176		
Total	59	0.219036476			

	Coefficients	*Standard Error*	*t Stat*	*P-value*	*Lower 95%*	*Upper 95%*
Intercept	-0.00633	0.007522567	-0.842	0.403246	-0.02139	0.008724
SP 500	0.967473	0.292113539	3.311977	0.001598	0.382745	1.552202

SUMMARY OUTPUT TATA

Regression Statistics	
Multiple R	0.477387
R Square	0.227898
Adjusted R Square	0.214586
Standard Error	0.130386
Observations	60

ANOVA

	df	*SS*	*MS*	*F*	*Significance F*
Regression	1	0.291044447	0.291044	17.11966	0.000115
Residual	58	0.986034863	0.017001		
Total	59	1.27707931			

	Coefficients	*Standard Error*	*t Stat*	*P-value*	*Lower 95%*	*Upper 95%*
Intercept	0.003951	0.01740474	0.226996	0.821225	-0.03089	0.03879
SP 500	2.796409	0.675854453	4.137591	0.000115	1.44354	4.149278

11.74

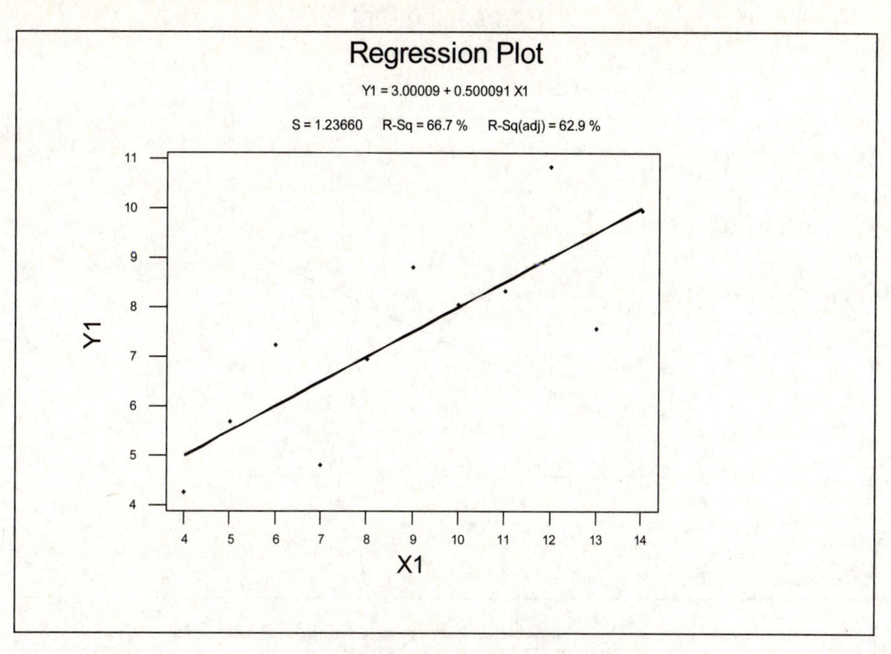

Regression Plot
Y1 = 3.00009 + 0.500091 X1
S = 1.23660 R-Sq = 66.7 % R-Sq(adj) = 62.9 %
Y1
X1

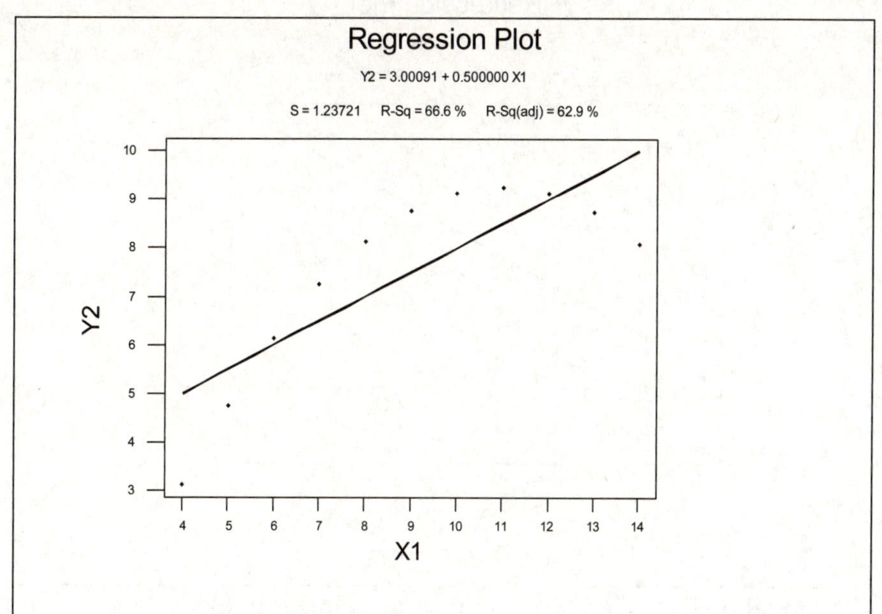

Regression Plot
Y2 = 3.00091 + 0.500000 X1
S = 1.23721 R-Sq = 66.6 % R-Sq(adj) = 62.9 %
Y2
X1

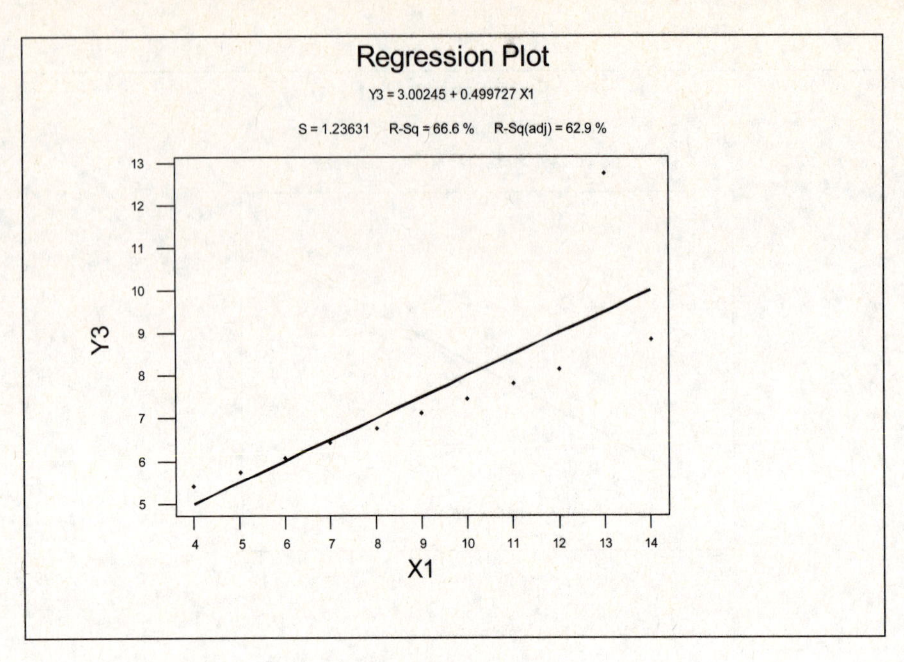

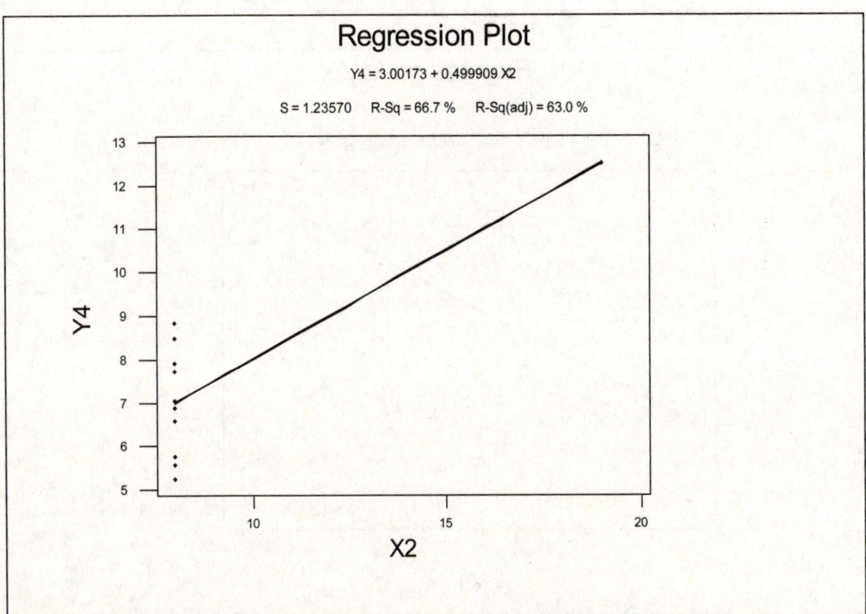

The model of Y1 = f(X1) is a good fit for a linear model
The model of Y2 = f(X1) is a non-linear model
The model of Y3 = f(X1) has a significant outlier at the largest value of X1
The model of Y4 = f(X1) has only two values of the independent variable

11.76 Two random variables are positively correlated if low values of one are associated with low values of the other and high values of one are associated with high values of the other

a. Total consumption expenditures are positively correlated with disposable income
b. Price of a good or service are negatively related with the quantity sold
c. The price of peanut butter and the sales of wrist watches are uncorrelated

11.78 $H_o: \rho = 0, H_1: \rho > 0$, $t = \frac{r\sqrt{(n-2)}}{\sqrt{(1-r^2)}} = t = \frac{.37\sqrt{51}}{\sqrt{(1-.37^2)}} = 2.844$

Therefore, reject H_0 at the .5% level since t = 2.844 > 2.666 ≈ $t_{51,.005}$

11.80 $H_o: \rho = 0, H_1: \rho > 0$, $t = \frac{r\sqrt{(n-2)}}{\sqrt{(1-r^2)}} = t = \frac{.293\sqrt{64}}{\sqrt{(1-.293^2)}} = 2.452$

Therefore, reject H_0 at the 1% level since t = 2.452 > 2.39 ≈ $t_{60,.01}$

11.82 To show this, let x = $\bar{x}$ for the regression of y on x, y = $b_0 + b_1x$

$\hat{y} = b_0 + b_1\bar{x} = \bar{y} - b\bar{x} + b\bar{x} = \bar{y}$

11.84 a. For a one unit change in the inflation rate, we estimate that the actual spot rate will change by .7916 units.

b. R^2 = 9.7%. 9.7% of the variation in the actual spot rate can be explained by the variations in the spot rate predicted by the inflation rate.

c. $H_o: \beta = 0, H_1: \beta > 0$, $t = \frac{.7916}{.2759} = 2.8692$,

Reject H_0 at the .5% level since t = 2.8692 > 2.66 = $t_{77,.005}$

d. $H_o: \beta = 0, H_1: \beta \neq 0$, $t = \frac{.7916 - 1}{.2759} = -.7553$,

Do not reject H_0 at any common level

11.86 a. For each unit increase in the diagnostic statistcs test, we estimate that the final student score at the end of the course will increase by .2875 points.

b. 11.58% of the variation in the final student score can be explained by the variation in the diagnostic statistics test

c. The two methods are 1) the test of the significance of the population regression slope coefficient (β) and 2) the test of the significance of the population correlation coefficient (ρ)

1) $H_o: \beta = 0, H_1: \beta > 0$, $t = \frac{.2875}{.04566} = 6.2965$

Therefore, reject H_0 at any common level of alpha

2) $H_o: \rho = 0, H_1: \rho > 0$, $r = \sqrt{R^2} = \sqrt{.1158} = .3403$

$t = \frac{r\sqrt{(n-2)}}{\sqrt{(1-r^2)}} = t = \frac{.3403\sqrt{304}}{\sqrt{(1-.3403^2)}} = 6.3098$, Reject H_0 at any common level

11.88 a. $R^2 = 1 - \frac{SSE}{SST} = 1 - \frac{204}{268} = .2388$. 23.88% of the variation in the dependent variable can be explained by the variation in the independent variable.

b. $\sum (x_i - \bar{x})^2 = \frac{SST(R^2)}{b^2} = \frac{268(.2388)}{(1.3)^2} = 37.8689$

$s^2_e = 204/23 = 8.8696$ $s^2_b = 8.8696/37.8689 = .2342$

$H_o: \beta = 0, H_1: \beta \neq 0$, $t = \frac{1.3}{\sqrt{.2342}} = 2.6863$

Therefore, reject H_0 at the 5% level since t = 2.6863 > 2.069 = $t_{23,.025}$

c. $1.3 \pm 2.069\sqrt{.2342}$, the interval runs from .2987 up to 2.3013

11.90 If a linear regression was estimated the slope would be negative and indicate that increased fertilizer reduces yield, which is silly.

11.92 a. Relationships are shown below in the correlation matrix with graphical plots to follow

Correlations: deaths, vehwt, impcars, lghttrks, carage

```
            deaths    vehwt  impcars lghttrks
vehwt        0.244
             0.091

impcars     -0.284   -0.943
             0.048    0.000

lghttrks     0.726    0.157   -0.175
             0.000    0.282    0.228

carage      -0.422    0.123    0.011   -0.329
             0.003    0.400    0.943    0.021

Cell Contents: Pearson correlation
               P-Value
```

Unusual data points include outliers of .55 crash deaths. This data point is much higher than expected given the levels of the independent variables.
Graphical plots (a) and Regression analyses (b):

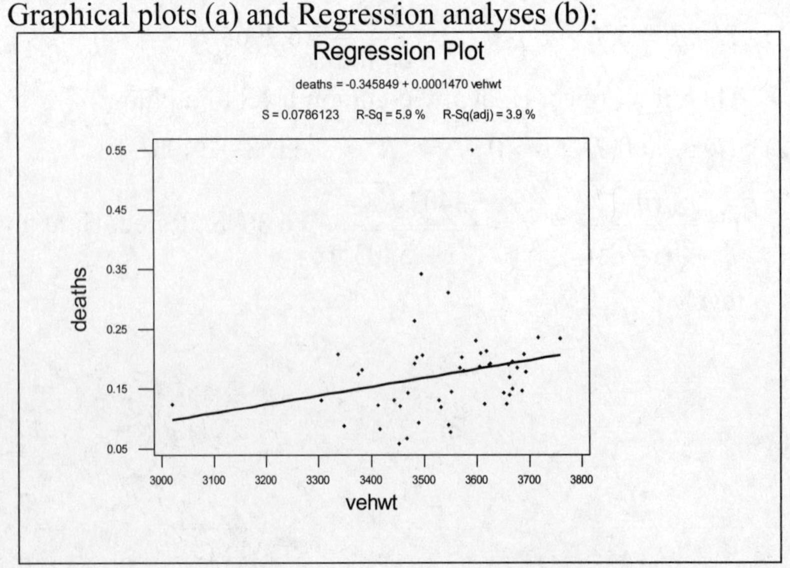

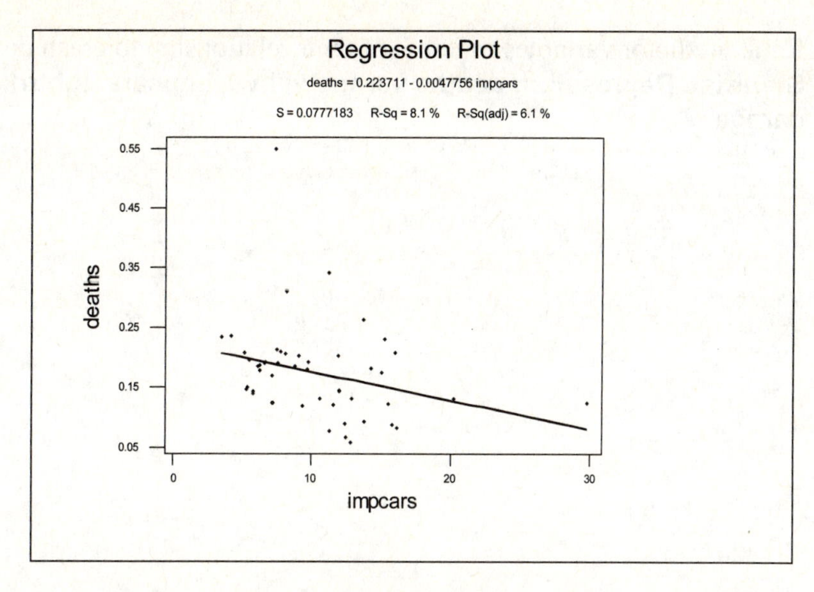

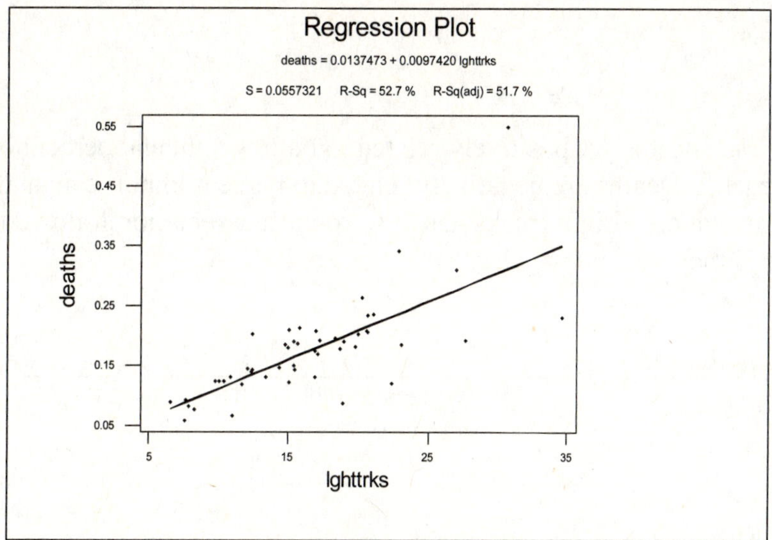

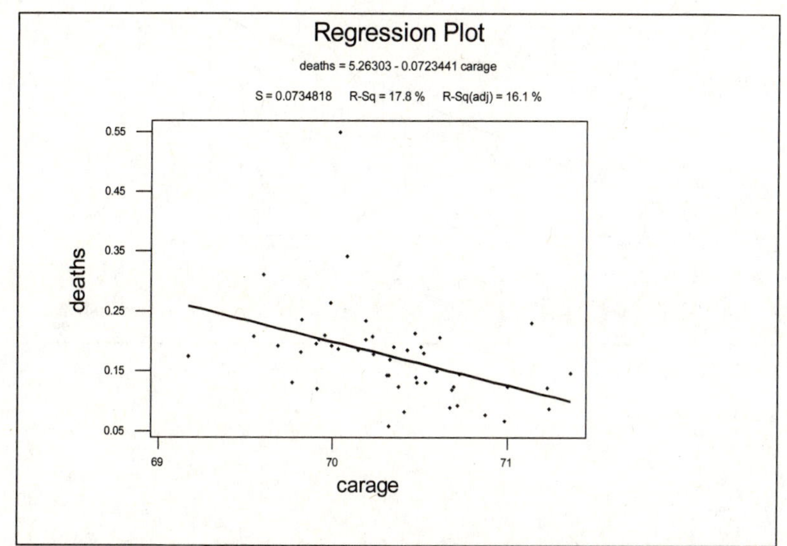

Light trucks has the strongest linear association (52.7%) followed by age (17.8%), then imported cars (8.1%) and then vehicle weight (5.9%).

c. Rank predictor variables in terms of their relationship to crash deaths

Stepwise Regression: deaths versus vehwt, impcars, lghttrks, carage

```
  Alpha-to-Enter: 0.15  Alpha-to-Remove: 0.15
 Response is  deaths  on  4 predictors, with N =   49

   Step             1         2         3
Constant      0.01375   2.50631   2.55472

lghttrks       0.0097    0.0088    0.0083
T-Value          7.24      6.39      6.02
P-Value         0.000     0.000     0.000

carage                   -0.035    -0.041
T-Value                   -2.00     -2.34
P-Value                   0.052     0.024

vehwt                             0.00011
T-Value                              1.80
P-Value                             0.079

S              0.0557    0.0540    0.0528
R-Sq            52.73     56.49     59.42
R-Sq(adj)       51.72     54.60     56.71
C-p               6.3       4.2       3.1
```

Crash deaths are positively related to both weight and percent of light trucks. Deaths are negatively related to percent import cars and the age of the vehicle. Light trucks has the strongest association followed by age and then vehicle weight.

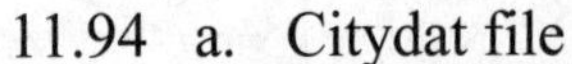

11.94 a. Citydat file

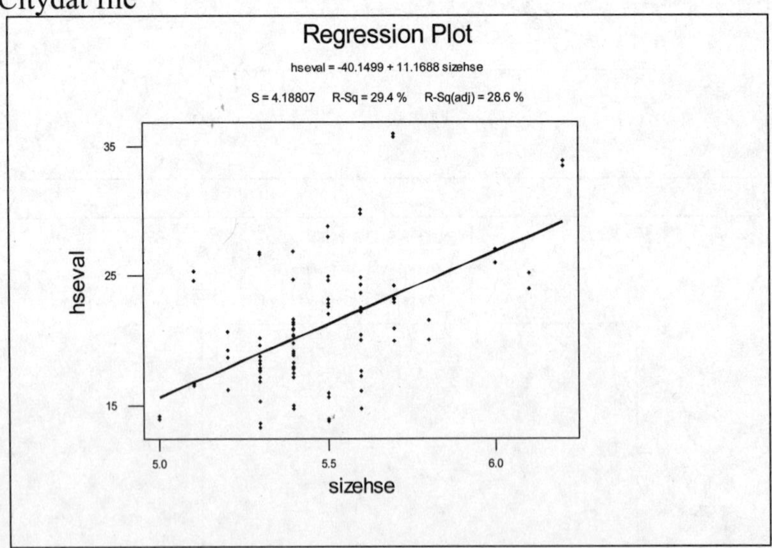

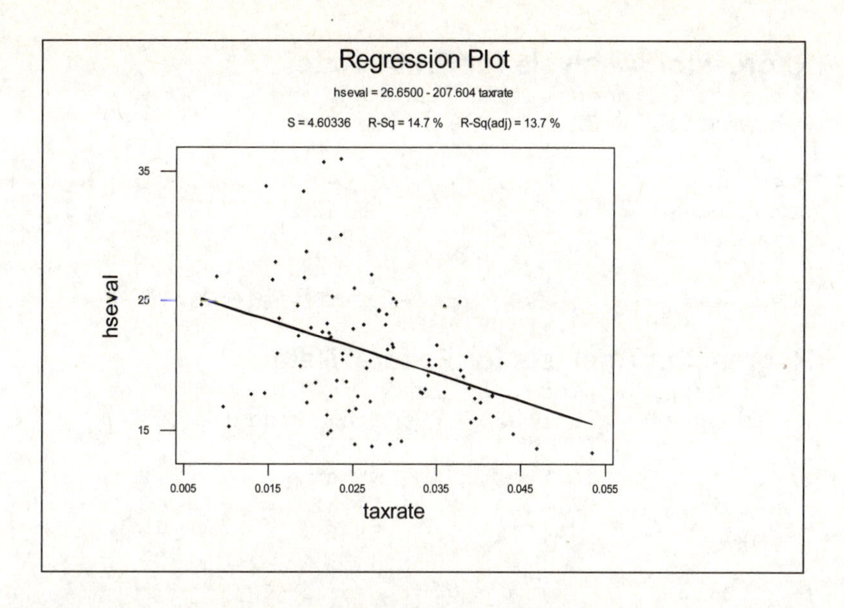

b. **Regression Analysis: hseval versus sizehse**

```
The regression equation is
hseval = - 40.1 + 11.2 sizehse
Predictor        Coef      SE Coef          T        P
Constant       -40.15        10.11      -3.97    0.000
sizehse        11.169        1.844       6.06    0.000

S = 4.188         R-Sq = 29.4%       R-Sq(adj) = 28.6%

Analysis of Variance
Source            DF         SS          MS          F        P
Regression         1     643.12      643.12      36.67    0.000
Residual Error    88    1543.51       17.54
Total             89    2186.63
```

Regression Analysis: hseval versus taxrate

```
The regression equation is
hseval = 26.6 - 208 taxrate
Predictor        Coef      SE Coef          T        P
Constant       26.650        1.521      17.52    0.000
taxrate       -207.60        53.27      -3.90    0.000

S = 4.603         R-Sq = 14.7%       R-Sq(adj) = 13.7%
Analysis of Variance
Source            DF         SS          MS          F        P
Regression         1     321.83      321.83      15.19    0.000
Residual Error    88    1864.80       21.19
Total             89    2186.63
```

Size of house is a stronger predictor than is the taxrate.

c. Whether tax rates are lowered or not does not have as large an impact as does the size of the house on the evaluation.

11.96 a. **Regression Analysis for Prime Rate**

The regression equation is
Residential = 224 + 7.79 Prime Rate

Predictor	Coef	StDev	T	P
Constant	223.59	18.42	12.14	0.000
Prime Ra	7.786	2.340	3.33	0.001

S = 122.1 R-Sq = 4.5% R-Sq(adj) = 4.1%

Regression Analysis for Federal Rate

The regression equation is
Residential = 264 + 2.81 Federal Funds Rate

Predictor	Coef	StDev	T	P
Constant	264.34	14.32	18.46	0.000
Federal	2.814	2.287	1.23	0.220

S = 124.6 R-Sq = 0.6% R-Sq(adj) = 0.2%

The $r^2 = 0.045$ for the regression with the prime rate is higher than the $r^2 = 0.006$ for the regression with the federal rate, so the regression with the prime rate explains more of the variation with the model. Also, the coefficient for the federal rate could statistically be equal to zero, i.e. the P-value is greater than 0.05.

b. Prime rate: $(3.19, 12.41)$, Federal rate: $(-1.68, 7.33)$

c. Prime rate 239.16, Federal rate 269.97

d. Prime rate: $(210.91, 267.41)$, Federal rate: $(248.57, 291.37)$

Chapter 12:

Multiple Regression

12.2 Given the following estimated linear model: $\hat{y} = 10 + 5x_1 + 4x_2 + 2x_3$

a. $\hat{y} = 10 + 5(20) + 4(11) + 2(10) = 174$
b. $\hat{y} = 10 + 5(15) + 4(14) + 2(20) = 181$
c. $\hat{y} = 10 + 5(35) + 4(19) + 2(25) = 311$
d. $\hat{y} = 10 + 5(10) + 4(17) + 2(30) = 188$

12.4 Given the following estimated linear model: $\hat{y} = 10 + 2x_1 + 12x_2 + 8x_3$

a. $\hat{y}$ increases by 8
b. $\hat{y}$ increases by 8
c. $\hat{y}$ increases by 24
d. $\hat{y}$ increases by 8

12.6 The estimated regression slope coefficients are interpreted as follows:
$b_1 = .661$: All else equal, an increase in the plane's top speed by one mph will increase the expected number of hours in the design effort by an estimated .661 million or 661 thousand worker-hours.
$b_2 = .065$: All else equal, an increase in the plane's weight by one ton will increase the expected number of hours in the design effort by an estimated .065 million or 65 thousand worker-hours
$b_3 = -.018$: All else equal, an increase in the percentage of parts in common with other models will result in a decrease in the expected number of hours in the design effort by an estimated .018 million or 18 thousand worker-hours

12.8 a. $b_1 = .052$: All else equal, an increase of one hundred dollars in weekly income results in an estimated .052 quarts per week increase in milk consumption. $b_2 = 1.14$: All else equal, an increase in family size by one person will result in an estimated increase in milk consumption by 1.14 quarts per week.

b. The intercept term b_0 of -.025 is the estimated milk consumption of quarts of milk per week given that the family's weekly income is 0 dollars and there are 0 members in the family. This is likely extrapolating beyond the observed data series and is not a useful interpretation.

12.10 Compute the slope coefficients for the model: $\hat{y}_i = b_0 + b_1 x_{1i} + b_2 x_{2i}$

Given that $b_1 = \frac{s_y(r_{x_1y} - r_{x_1x_2} r_{x_2y})}{s_{x_1}(1 - r^2_{x_1x_2})}$, $b_2 = \frac{s_y(r_{x_2y} - r_{x_1x_2} r_{x_1y})}{s_{x_2}(1 - r^2_{x_1x_2})}$

a. $b_1 = \frac{400(.60 - (.50)(.70))}{200(1 - .50^2)} = 2.000$, $b_2 = \frac{400(.70 - (.50)(.60))}{200(1 - .50^2)} = 3.200$

b. $b_1 = \frac{400(-.60 - (-.50)(.70))}{200(1 - (-.50)^2)} = -.667$,

$b_2 = \frac{400(.70 - (-.50)(-.60))}{200(1 - (-.50)^2)} = 1.067$

c. $b_1 = \frac{400(.40 - (.80)(.45))}{200(1 - (.80)^2)} = .083$, $b_2 = \frac{400(.45 - (.80)(.40))}{200(1 - (.80)^2)} = .271$

d. $b_1 = \frac{400(.60 - (-.60)(-.50))}{200(1 - (-.60)^2)} = .9375$,

$b_1 = \frac{400(-.50 - (-.60)(.60))}{200(1 - (-.60)^2)} = -.4375$

12.12 a. Electricity sales as a function of number of customers and price

Regression Analysis: salesmw2 versus priclec2, numcust2

```
The regression equation is
salesmw2 = - 647363 + 19895 priclec2 + 2.35 numcust2
Predictor        Coef    SE Coef         T      P
Constant      -647363     291734     -2.22  0.030
priclec2        19895      22515      0.88  0.380
numcust2       2.3530     0.2233     10.54  0.000

S = 66399        R-Sq = 79.2%      R-Sq(adj) = 78.5%
Analysis of Variance
Source            DF          SS          MS         F       P
Regression         2 1.02480E+12 5.12400E+11    116.22   0.000
Residual Error    61 2.68939E+11  4408828732
Total             63 1.29374E+12
```

All else equal, for every one unit increase in the price of electricity, we estimate that sales will increase by 19895 mwh. Note that this estimated coefficient is not significantly different from zero (p-value = .380).
All else equal, for every additional residential customer who uses electricity in the heating of their home, we estimate that sales will increase by 2.353 mwh.

b. Electricity sales as a function of number of customers

Regression Analysis: salesmw2 versus numcust2

```
The regression equation is
salesmw2 = - 410202 + 2.20 numcust2
Predictor        Coef    SE Coef         T      P
Constant      -410202     114132     -3.59  0.001
numcust2       2.2027     0.1445     15.25  0.000

S = 66282        R-Sq = 78.9%      R-Sq(adj) = 78.6%
Analysis of Variance
Source            DF          SS          MS         F       P
Regression         1 1.02136E+12 1.02136E+12    232.48   0.000
Residual Error    62 2.72381E+11  4393240914
Total             63 1.29374E+12
```

An additional residential customer will add 2.2027 mwh to electricity sales.
The two models have roughly equivalent explanatory power; therefore, adding price as a variable does not add a significant amount of explanatory power to the model. There appears to be a problem of high correlation between the independent variables of price and customers.

c. **Regression Analysis: salesmw2 versus priclec2, degrday2**

```
The regression equation is
salesmw2 = 2312260 - 165275 priclec2 + 56.1 degrday2
Predictor        Coef     SE Coef         T        P
Constant      2312260      148794     15.54    0.000
priclec2      -165275       24809     -6.66    0.000
degrday2        56.06       60.37      0.93    0.357

S = 110725       R-Sq = 42.2%      R-Sq(adj) = 40.3%
Analysis of Variance
Source            DF          SS          MS         F        P
Regression         2 5.45875E+11 2.72938E+11     22.26    0.000
Residual Error    61 7.47863E+11 12260053296
Total             63 1.29374E+12
```

All else equal, an increase in the price of electricity will reduce electricity sales by 165,275 mwh.
All else equal, an increase in the degree days (departure from normal weather) by one unit will increase electricity sales by 56.06 mwh.
Note that the coefficient on the price variable is now negative, as expected, and it is significantly different from zero (p-value = .000)

d. **Regression Analysis: salesmw2 versus Yd872, degrday2**

```
The regression equation is
salesmw2 = 293949 + 326 Yd872 + 58.4 degrday2
Predictor        Coef     SE Coef         T        P
Constant       293949       67939      4.33    0.000
Yd872          325.85       21.30     15.29    0.000
degrday2        58.36       35.79      1.63    0.108

S = 66187        R-Sq = 79.3%      R-Sq(adj) = 78.7%
Analysis of Variance
Source            DF          SS          MS         F        P
Regression         2 1.02652E+12 5.13259E+11    117.16    0.000
Residual Error    61 2.67221E+11  4380674677
Total             63 1.29374E+12
```

All else equal, an increase in personal disposable income by one unit will increase electricity sales by 325.85 mwh.
All else equal, an increase in degree days by one unit will increase electricity sales by 58.36 mwh.

12.1 a. Horsepower as a function of weight, cubic inches of displacement

Regression Analysis: horspwr versus weight, displace

```
The regression equation is
horspwr = 23.5 + 0.0154 weight + 0.157 displace
151 cases used 4 cases contain missing values
Predictor        Coef     SE Coef        T        P        VIF
Constant       23.496       7.341     3.20    0.002
weight       0.015432    0.004538     3.40    0.001        6.0
displace      0.15667     0.03746     4.18    0.000        6.0

S = 13.64       R-Sq = 69.2%      R-Sq(adj) = 68.8%
Analysis of Variance
Source            DF          SS         MS        F        P
Regression         2       61929      30964   166.33    0.000
Residual Error   148       27551        186
Total            150       89480
```

All else equal, a 100 pound increase in the weight of the car is associated with a 1.54 increase in horsepower of the auto.

All else equal, a 10 cubic inch increase in the displacement of the engine is associated with a 1.57 increase in the horsepower of the auto.

b. Horsepower as a function of weight, displacement, number of cylinders

Regression Analysis: horspwr versus weight, displace, cylinder

```
The regression equation is
horspwr = 16.7 + 0.0163 weight + 0.105 displace + 2.57 cylinder
151 cases used 4 cases contain missing values
Predictor        Coef     SE Coef        T        P        VIF
Constant       16.703       9.449     1.77    0.079
weight       0.016261    0.004592     3.54    0.001        6.2
displace      0.10527     0.05859     1.80    0.074       14.8
cylinder        2.574       2.258     1.14    0.256        7.8

S = 13.63       R-Sq = 69.5%      R-Sq(adj) = 68.9%
Analysis of Variance
Source            DF          SS         MS        F        P
Regression         3       62170      20723   111.55    0.000
Residual Error   147       27310        186
Total            150       89480
```

All else equal, a 100 pound increase in the weight of the car is associated with a 1.63 increase in horsepower of the auto.

All else equal, a 10 cubic inch increase in the displacement of the engine is associated with a 1.05 increase in the horsepower of the auto.

All else equal, one additional cylinder in the engine is associated with a 2.57 increase in the horsepower of the auto.

Note that adding the independent variable number of cylinders has not added to the explanatory power of the model. R square has increased marginally. Engine displacement is no longer significant at the .05 level (p-value of .074) and the estimated regression slope coefficient on the number of cylinders is not significantly different from zero. This is due to the strong correlation that exists between cubic inches of engine displacement and the number of cylinders.

c. Horsepower as a function of weight, displacement and fuel mileage

Regression Analysis: horspwr versus weight, displace, milpgal

```
The regression equation is
horspwr = 93.6 + 0.00203 weight + 0.165 displace - 1.24 milpgal
150 cases used 5 cases contain missing values
Predictor        Coef      SE Coef         T        P         VIF
Constant        93.57        15.33      6.11    0.000
weight       0.002031     0.004879      0.42    0.678         8.3
displace      0.16475      0.03475      4.74    0.000         6.1
milpgal       -1.2392       0.2474     -5.01    0.000         3.1

S = 12.55        R-Sq = 74.2%      R-Sq(adj) = 73.6%
Analysis of Variance
Source             DF           SS        MS         F         P
Regression          3        66042     22014    139.77     0.000
Residual Error    146        22994       157
Total             149        89036
```

All else equal, a 100 pound increase in the weight of the car is associated with a .203 increase in horsepower of the auto.
All else equal, a 10 cubic inch increase in the displacement of the engine is associated with a 1.6475 increase in the horsepower of the auto.
All else equal, an increase in the fuel mileage of the vehicle by 1 mile per gallon is associated with a reduction in horsepower of 1.2392.
Note that the negative coefficient on fuel mileage indicates the trade-off that is expected between horsepower and fuel mileage. The displacement variable is significantly positive, as expected, however, the weight variable is no longer significant. Again, one would expect high correlation among the independent variables.

d. Horsepower as a function of weight, displacement, mpg and price

Regression Analysis: horspwr versus weight, displace, milpgal, price

```
The regression equation is
horspwr = 98.1 - 0.00032 weight + 0.175 displace - 1.32 milpgal
+0.000138 price
150 cases used 5 cases contain missing values
Predictor        Coef      SE Coef         T        P         VIF
Constant        98.14        16.05      6.11    0.000
weight      -0.000324     0.005462     -0.06    0.953        10.3
displace      0.17533      0.03647      4.81    0.000         6.8
milpgal       -1.3194       0.2613     -5.05    0.000         3.5
price       0.0001379    0.0001438      0.96    0.339         1.3

S = 12.55        R-Sq = 74.3%      R-Sq(adj) = 73.6%
Analysis of Variance
Source             DF           SS        MS         F         P
Regression          4        66187     16547    105.00     0.000
Residual Error    145        22849       158
Total             149        89036
```

Engine displacement has a significant positive impact on horsepower, fuel mileage is negatively related to horsepower and price is not significant.

e. Explanatory power has marginally increased from the first model to the last. The estimated coefficient on price is not significantly different from zero. Displacement and fuel mileage have the expected signs. The coefficient on weight has the wrong sign; however, it is not significantly different from zero (p-value of .953).

12.16 A regression analysis has produced the following Analysis of Variance table

Given that SST = SSR + SSE, $s^2_e = \frac{SSE}{n-k-1}$, $R^2 = \frac{SSR}{SST} = 1 - \frac{SSE}{SST}$,

$\bar{R}^2 = 1 - \frac{SSE/(n-k-1)}{SST/(n-1)}$

a. $SSE = 2500,\ s^2_e = \frac{2500}{32-2-1} = 86.207,\ s_e = 9.2848$

b. SST = SSR + SSE = 7,000 + 2,500 = 9,500

c. $R^2 = \frac{7000}{9500} = 1 - \frac{2500}{9500} = .7368,\ \bar{R}^2 = 1 - \frac{2500/(29)}{9500/(31)} = .7086$

12.18 A regression analysis has produced the following Analysis of Variance table

Given that SST = SSR + SSE, $s^2_e = \frac{SSE}{n-k-1}$, $R^2 = \frac{SSR}{SST} = 1 - \frac{SSE}{SST}$,

$\bar{R}^2 = 1 - \frac{SSE/(n-k-1)}{SST/(n-1)}$

a. $SSE = 15{,}000,\ s^2_e = \frac{15{,}000}{206-5-1} = 75.0,\ s_e = 8.660$

b. SST = SSR + SSE = 80,000 + 15,000 = 95,000

c. $R^2 = \frac{80{,}000}{95{,}000} = 1 - \frac{15{,}000}{95{,}000} = .8421,\ \bar{R}^2 = 1 - \frac{10{,}000/(45)}{50{,}000/(49)} = .7822$

12.20 a. $R^2 = \frac{88.2}{162.1} = .5441$, therefore, 54.41% of the variability in milk consumption can be explained by the variations in weekly income and family size.

b. $\bar{R}^2 = 1 - \frac{73.9/(30-3)}{162.1/29} = .5103$

c. $R = \sqrt{.5441} = .7376$. This is the sample correlation between observed and predicted values of milk consumption.

12.22 a. **Regression Analysis: Y profit versus X2 offices**

```
The regression equation is
Y profit = 1.55 -0.000120 X2  offices
Predictor           Coef      SE Coef          T        P
Constant          1.5460       0.1048      14.75    0.000
X2  offi    -0.00012033   0.00001434      -8.39    0.000

S = 0.07049    R-Sq = 75.4%     R-Sq(adj) = 74.3%
Analysis of Variance
Source            DF          SS          MS         F        P
Regression         1     0.34973     0.34973     70.38    0.000
Residual Error    23     0.11429     0.00497
Total             24     0.46402
```

b. **Regression Analysis: X1 revenue versus X2 offices**

```
The regression equation is
X1 revenue = - 0.078 +0.000543 X2  offices

Predictor          Coef       SE Coef         T        P
Constant        -0.0781        0.2975     -0.26    0.795
X2  offi    0.00054280    0.00004070     13.34    0.000

S = 0.2000      R-Sq = 88.5%     R-Sq(adj) = 88.1%
Analysis of Variance
Source            DF          SS         MS        F        P
Regression         1      7.1166     7.1166   177.84    0.000
Residual Error    23      0.9204     0.0400
Total             24      8.0370
```

c. **Regression Analysis: Y profit versus X1 revenue**

```
The regression equation is
Y profit = 1.33 - 0.169 X1 revenue
Predictor          Coef       SE Coef         T        P
Constant         1.3262        0.1386      9.57    0.000
X1 reven       -0.16913       0.03559     -4.75    0.000

S = 0.1009      R-Sq = 49.5%     R-Sq(adj) = 47.4%
Analysis of Variance
Source            DF          SS         MS        F        P
Regression         1     0.22990    0.22990    22.59    0.000
Residual Error    23     0.23412    0.01018
Total             24     0.46402
```

d. **Regression Analysis: X2 offices versus X1 revenue**

```
The regression equation is
X2  offices = 957 + 1631 X1 revenue
Predictor          Coef       SE Coef         T        P
Constant          956.9         476.5      2.01    0.057
X1 reven         1631.3         122.3     13.34    0.000

S = 346.8       R-Sq = 88.5%     R-Sq(adj) = 88.1%
Analysis of Variance
Source            DF          SS         MS        F        P
Regression         1    21388013   21388013   177.84    0.000
Residual Error    23     2766147     120267
Total             24    24154159
```

12.24 Given the following results where the numbers in parentheses are the sample standard error of the coefficient estimates

a. Compute two-sided 95% confidence intervals for the three regression slope coefficients

$b_j \pm t_{n-k-1,\alpha/2} s_{b_j}$

95% CI for x_1 = 6.8 ± 2.042 (3.1); .4698 up to 13.1302

95% CI for x_2 = 6.9 ± 2.042 (3.7); -6.4554 up to 14.4554

95% CI for x_3 = -7.2 ± 2.042 (3.2); -13.7344 up to -.6656

b. Test the hypothesis $H_0: \beta_j = 0, H_1: \beta_j > 0$

For x_1: $t = \frac{6.8}{3.1} = 2.194 \quad t_{30,.05/.01} = 1.697, 2.457$

Therefore, reject H_0 at the 5% level but not at the 1% level

For x_2: $t = \frac{6.9}{3.7} = 1.865 \quad t_{30,.05/.01} = 1.697, 2.457$

Therefore, reject H_0 at the 5% level but not at the 1% level

For x_3: $t = \frac{-7.2}{3.2} = -2.25 \quad t_{30,.05/.01} = 1.697, 2.457$

Therefore, do not reject H_0 at the 5% level nor the 1% level

12.26 Given the following results where the numbers in parentheses are the sample standard error of the coefficient estimates

a. Compute two-sided 95% confidence intervals for the three regression slope coefficients

$b_j \pm t_{n-k-1,\alpha/2} s_{b_j}$

95% CI for x_1 = 17.8 ± 2.042 (7.1); 3.3018 up to 32.2982
95% CI for x_2 = 26.9 ± 2.042 (13.7); -1.0754 up to 54.8754
95% CI for x_3 = -9.2 ± 2.042 (3.8); -16.9596 up to -1.44

b. Test the hypothesis $H_0: \beta_j = 0, H_1: \beta_j > 0$

For x_1: $t = \frac{17.8}{7.1} = 2.507 \quad t_{35,.05/.01} \approx 1.697, 2.457$

Therefore, reject H_0 at the 5% level but not at the 1% level

For x_2: $t = \frac{26.9}{13.7} = 1.964 \quad t_{35,.05/.01} \approx 1.697, 2.457$

Therefore, reject H_0 at the 5% level but not at the 1% level

For x_3: $t = \frac{-9.2}{3.8} = -2.421 \quad t_{35,.05/.01} \approx 1.697, 2.457$

Therefore, do not reject H_0 at either level

12.28 a. $H_0: \beta_1 = 0; H_1: \beta_1 > 0$

$t = \frac{.052}{.023} = 2.26$

$t_{27,.025/.01} = 2.052, 2.473$

Therefore, reject H_0 at the 2.5% level but not at the 1% level

b. $t_{27,.05/.025/.005} = 1.703, 2.052, 2.771$

90% CI: 1.14 ± 1.703(.35); .5439 up to 1.7361
95% CI: 1.14 ± 2.052(.35); .4218 up to 1.8582
99% CI: 1.14 ± 2.771(.35); .1701 up to 2.1099

12.30 a. $H_0: \beta_3 = 0, H_1: \beta_3 \neq 0$

$$t = \frac{-.000191}{.000446} = -.428$$

$t_{16,.10} = -1.337$

Therefore, do not reject H_0 at the 20% level

b. $H_0: \beta_1 = \beta_2 = \beta_3 = 0, H_1:$ At least one $\beta_i \neq 0, (i = 1,2,3)$

$$F = \frac{16}{3}\frac{.71}{1-.71} = 13.057, \; F_{3,16,.01} = 5.29$$

Therefore, reject H_0 at the 1% level

12.32 a. All else being equal, an extra $1 in mean per capita personal income leads to an expected extra $.04 of net revenue per capita from the lottery

b. $b_2 = .8772, s_{b_2} = .3107, n = 29, t_{24,.025} = 2.064$

95% CI: $.8772 \pm 2.064(.3107)$, .2359 up to 1.5185

c. $H_0: \beta_3 = 0, H_1: \beta_3 < 0$

$$t = \frac{-365.01}{263.88} = -1.383$$

$t_{24,.10/.05} = -1.318, -1.711$

Therefore, reject H_0 at the 10% level but not at the 5% level

12.34 a. $n = 19, b_1 = .2, s_{b_1} = .0092, t_{16,.025} = 2.12$

95% CI: $.2 \pm 2.12(.0092)$, .1805 up to .2195

b. $H_0: \beta_2 = 0, H_1: \beta_2 < 0, \; t = \frac{-.1}{.084} = -1.19$

$t_{16,.10} = -1.337$, Therefore, do not reject H_0 at the 10% level

12.36 a. $n = 39, b_5 = .0495, s_{b_1} = .01172, t_{30,.005} = 2.750$

99% CI: $.0495 \pm 2.750(.01172)$, .0173 up to .0817

b. $H_0: \beta_4 = 0, H_1: \beta_4 \neq 0$

$$t = \frac{.48122}{.77954} = .617$$

$t_{30,.10} = 1.31$

Therefore, do not reject H_0 at the 20% level

c. $H_0: \beta_7 = 0, H_1: \beta_7 \neq 0$

$$t = \frac{.00645}{.00306} = 2.108$$

$t_{30,.025/.01} = 2.042, 2.457$

Therefore, reject H_0 at the 5% level but not at the 2% level

12.38 a. SST = 3.881, SSR = 3.549, SSE = .332

$H_0: \beta_1 = \beta_2 = \beta_3 = 0, H_1:$ At least one $\beta_i \neq 0, (i = 1,2,3)$

$$F = \frac{3.549/3}{.332/23} = 81.955$$

$F_{3,23,.01} = 4.76$

Therefore, reject H_0 at the 1% level.

b. Analysis of Variance table:

Sources of variation	Sum of Squares	Degress of Freedom	Mean Squares	F-Ratio
Regressor	3.549	3	1.183	81.955
Error	.332	23	.014435	
Total	3.881	26		

12.40 a. SST = 162.1, SSR =88.2, SSE = 73.9

$H_0: \beta_1 = \beta_2 = 0, H_1:$ At least one $\beta_i \neq 0, (i = 1,2)$

$$F = \frac{88.2/2}{73.9/27} = 16.113, \quad F_{2,27,.01} = 5.49$$

Therefore, reject H_0 at the 1% level

b.

Sources of variation	Sum of Squares	Degress of Freedom	Mean Squares	F-Ratio
Regressor	88.2	2	44.10	16.113
Error	73.9	27	2.737	
Total	162.1	29		

12.42 $H_0: \beta_1 = \beta_2 = \beta_3 = \beta_4 = 0, H_1:$ At least one $\beta_i \neq 0, (i = 1,2,3,4)$

The test can be based directly on the coefficient of determination since

$$R^2 = \frac{SSR}{SST} = 1 - \frac{SSE}{SST}, \text{ and hence } \frac{R^2}{1-R^2} = \frac{SSR/SST}{SSE/SST} = \frac{SSR}{SSE} = F, \text{ and}$$

$$F = \frac{n-K-1}{K}\left(\frac{R^2}{1-R^2}\right), \quad F = \frac{24}{4}\frac{.51}{1-.51} = 6.2449, F_{4,24,.01} = 4.22. \text{ Therefore,}$$

reject H_0 at the 1% level

12.44 $H_0: \beta_1 = \beta_2 = 0, H_1:$ At least one $\beta_i \neq 0, (i = 1,2)$

$$R^2 = \frac{SSR}{SST} = 1 - \frac{SSE}{SST}, \text{ and hence } \frac{R^2}{1-R^2} = \frac{SSR/SST}{SSE/SST} = \frac{SSR}{SSE} = F, \text{ and}$$

$$F = \frac{n-K-1}{K}\left(\frac{R^2}{1-R^2}\right), \quad F = \frac{16}{2}\frac{.96+(2/16)}{1-.96} = 217, \quad F_{2,16,.01} = 6.23$$

Therefore, reject H_0 at the 1% level

12.46 $\dfrac{(SSE*-SSE)k_1}{SSE/(n-k-1)} = \dfrac{n-k-1}{k_1}\dfrac{(SSE*-SSE)/SST}{SSE/SST}$

$$= \frac{n-k-1}{k_1}\frac{1-R^2{*}-(1-R^2)}{1-R^2} = \frac{n-k-1}{k_1}\frac{R^2-R^2{*}}{1-R^2}$$

12.48 a. $\bar{R}^2 = 1-\dfrac{SSE/(n-k-1)}{SST/(n-1)} = 1-\dfrac{n-1}{n-k-1}(1-R^2) = \dfrac{n-1}{n-k-1}R^2 - \dfrac{k}{n-k-1}$

$$= \frac{(n-1)R^2-k}{n-k-1}$$

b. Since $\bar{R}^2 = \dfrac{(n-1)R^2-k}{n-k-1}$, then $R^2 = \dfrac{(n-k-1)\bar{R}^2+k}{n-1}$

c. $\dfrac{SSR/k}{SSE/(n-k-1)} = \dfrac{n-k-1}{k}\dfrac{SSR/SST}{SSE/SST} = \dfrac{n-k-1}{k}\dfrac{R^2}{1-R^2} =$

$$\frac{n-k-1}{k}\frac{[(n-k-1)\bar{R}^2+k]/(n-1)}{[n-1-(n-k-1)\bar{R}^2-k]/(n-1)} = \frac{n-k-1}{k}\frac{(n-k-1)\bar{R}^2+k}{(n-k-1)(1-\bar{R}^2)}$$

$$= \frac{n-k-1}{k}\frac{\bar{R}^2+k}{(1-\bar{R}^2)}$$

12.50 $\hat{Y} = 7.35+.653(20)-1.345(10)+.613(6) = 10.638$ pounds

12.52 $\hat{Y} = 2.0+.661(1)+.065(7)-.018(50) = 2.216$ million worker hours

12.54 Compute values of y_i when x_i = 1, 2, 4, 6, 8, 10

Xi	1	2	4	6	8	10
$y_i = 4x^{1.5}$	4	11.3137	32	58.7878	90.5097	126.4611
$y_i = 1+2x_i+2x_i^2$	5	13	41	85	145	221

12.56 Compute values of y_i when x_i = 1, 2, 4, 6, 8, 10

Xi	1	2	4	6	8	10
$y_i = 4x^{1.5}$	4	11.3137	32	58.7878	90.5097	126.4611
$y_i = 1+2x_i+1.7x_i^2$	4.7	11.8	36.2	74.2	125.8	191

12.58 There are many possible answers. Relationships that can be approximated by a non-linear quadratic model include many supply functions, production functions and cost functions including average cost versus the number of units produced.

12.60 a. All else equal, 1% increase in annual consumption expenditures will be associated with a 1.1556% increase in expenditures on vacation travel. All else equal, a 1% increase in the size of the household will be associated with a .4408% decrease in expenditures on vacation travel.

b. 16.8% of the variation in vacation travel expenditures can be explained by the variations in the log of total consumption expenditures and log of the number of members in the household

c. $1.1556 \pm 1.96(.0546) = 1.049$ up to 1.2626

d. $H_0: \beta_2 = 0, H_1: \beta_2 < 0$, $t = \frac{-.4408}{.0490} = -8.996$,

Therefore, reject H_0 at the 1% level

12.62 a. All else equal, a 1% increase in the price of beef will be associated with a decrease of .529% in the tons of beef consumed annually in the U.S.

b. All else equal, a 1% increase in the price of pork will be associated with an increase of .217% in the tons of beef consumed annually in the U.S.

c. $H_0: \beta_4 = 0, H_1: \beta_4 > 0$, $t = \frac{.416}{.163} = 2.552$, $t_{25,.01} = 2.485$,

Therefore, reject H_0 at the 1% level

d. $H_0: \beta_1 = \beta_2 = \beta_3 = \beta_4 = 0, H_1:$ At least one $\beta_i \neq 0, (i = 1,2,3,4)$

$$F = \frac{n-k-1}{k} \frac{R^2}{1-R^2} = \frac{25}{4} \frac{.683}{1-.683} = 13.466, F_{4,25,.01} = 4.18.$$

Therefore, reject H_0 at the 1% level

e. If an important independent variable has been omitted, there may be specification bias. The regression coefficients produced for the misspecified model would be misleading.

12.64 a. Coefficients for exponential models can be estimated by taking the logarithm of both sides of the multiple regression model to obtain an equation that is linear in the logarithms of the variables.

$\log(Y) = \log(\beta_0) + \beta_1 \log(X_1) + \beta_2 \log(X_2) + \beta_3 \log(X_3) + \beta_4 (\log(X_4) + \log(\varepsilon)$

Substituting in the restrictions on the coefficients: $\beta_1 + \beta_2 = 1, \beta_2 = 1 - \beta_1$, $\beta_3 + \beta_4 = 1, \beta_4 = 1 - \beta_3$

$\log(Y) = \log(\beta_0) + \beta_1 \log(X_1) + [1 - \beta_1]\log(X_2) + \beta_3 \log(X_3) + [1 - \beta_3](\log(X_4) + \log(\varepsilon)$

Simplify algebraically and estimate the coefficients. The coefficient β_2 can be found by subtracting β_1 from 1.0. Likewise the coefficient β_4 can be found by subtracting β_3 from 1.0.

b. Constant elasticity for Y versus X_4 is the regression slope coefficient on the X_4 term of the logarithm model.

12.66 **Results for: GermanImports.xls**
Regression Analysis: LogYt versus LogX1t, LogX2t

```
The regression equation is
LogYt = - 4.07 + 1.36 LogX1t + 0.101 LogX2t
Predictor        Coef     SE Coef        T        P      VIF
Constant      -4.0709      0.3100   -13.13    0.000
LogX1t        1.35935     0.03005    45.23    0.000      4.9
LogX2t        0.10094     0.05715     1.77    0.088      4.9

S = 0.04758     R-Sq = 99.7%     R-Sq(adj) = 99.7%

Analysis of Variance
Source            DF        SS        MS        F        P
Regression         2    21.345    10.673  4715.32    0.000
Residual Error    28     0.063     0.002
Total             30    21.409

Source       DF     Seq SS
LogX1t        1     21.338
LogX2t        1      0.007
```

12.68 What is the model constant when the dummy variable equals 1

a. $\hat{y} = 5.78 + 4.87x_1$

b. $\hat{y} = 1.15 + 9.51x_1$

c. $\hat{y} = 13.67 + 8.98x_1$

12.70 a. All else being equal, expected selling price is higher by \$3,219 if condo has a fireplace.

b. All else being equal, expected selling price is higher by \$2,005 if condo has brick siding.

c. 95% CI: $3219 \pm 1.96(947) = \$1,362.88$ up to \$5,075.12

d. $H_0 : \beta_5 = 0, H_1 : \beta_5 > 0$, $t = \dfrac{2005}{768} = 2.611$, $t_{809,.005} = 2.576$

Therefore, reject H_0 at the .5% level

12.72 35.6% of the variation in overall performance in law school can be explained by the variation in undergraduate gpa, scores on the LSATs and whether the student's letter of recommendation are unusually strong. The overall model is significant since we can reject the null hypothesis that the model has no explanatory power in favor of the alternative hypothesis that the model has significant explanatory power. The individual regression coefficients that are significantly different than zero include the scores on the LSAT and whether the student's letters of recommendation were unusually strong. The coefficient on undergraduate gpa was not found to be significant at the 5% level.

12.74 a. All else equal, the average rating of a course is 6.21 units higher if a visiting lecturer is brought in than if otherwise.

b. $H_0 : \beta_4 = 0, H_1 : \beta_4 > 0$, $t = \frac{6.21}{3.59} = 1.73$, $t_{20,.05} = 1.725$

Therefore, reject H_0 at the 5% level

c. 56.9% of the variation in the average course rating can be explained by the variation in the percentage of time spent in group discussions, the dollars spent on preparing the course materials, the dollars spent on food and drinks, and whether a guest lecturer is brought in.

$H_0 : \beta_1 = \beta_2 = \beta_3 = \beta_4 = 0, H_1 :$ At least one $\beta_i \neq 0, (i = 1,2,3,4)$

$$F = \frac{n-k-1}{k}\frac{R^2}{1-R^2} = \frac{20}{4}\frac{.569}{1-.569} = 6.6$$

$F_{4,20,.01} = 4.43$

Therefore, reject H_0 at the 1% level

d. $t_{20,.025} = 2.086$

95% CI: $.52 \pm 2.086(.21)$, .0819 up to .9581

12.76 **Results for: Student Performance.xls**

Regression Analysis: Y versus X1, X2, X3, X4, X5

```
The regression equation is
Y = 2.00 + 0.0099 X1 + 0.0763 X2 - 0.137 X3 + 0.064 X4 + 0.138 X5

Predictor        Coef    SE Coef       T       P    VIF
Constant        1.997      1.273    1.57   0.132
X1            0.00990    0.01654    0.60   0.556    1.3
X2            0.07629    0.05654    1.35   0.192    1.2
X3           -0.13652    0.06922   -1.97   0.062    1.1
X4             0.0636     0.2606    0.24   0.810    1.4
X5            0.13794    0.07521    1.83   0.081    1.1

S = 0.5416       R-Sq = 26.5%      R-Sq(adj) = 9.0%

Analysis of Variance
Source           DF        SS        MS        F       P
Regression        5    2.2165    0.4433     1.51   0.229
Residual Error   21    6.1598    0.2933
Total            26    8.3763
```

The model is not significant (p-value of the F-test = .229). The model only explains 26.5% of the variation in gpa with the hours spent studying, hours spent preparing for tests, hours spent in bars, whether or not students take notes or mark highlights when reading tests and the average number of credit hours taken per semester. The only independent variables that are marginally significant (10% level but not the 5% level) include number of hours spent in bars and average number of credit hours. The other independent variables are not significant at common levels of alpha.

12.78 Two variables are included as predictor variables. What is the effect on the estimated slope coefficients when these two variables have a correlation equal to

a. .78. A large correlation among the independent variables will lead to a high variance for the estimated slope coefficients and will tend to have a small student's t statistic. Use the rule of thumb $|r| > \frac{2}{\sqrt{n}}$ to determine if the correlation is 'large'.

b. .08. No correlation exists among the independent variables. No effect on the estimated slope coefficients.

c. .94. A large correlation among the independent variables will lead to a high variance for the estimated slope coefficients and will tend to have a small student's t statistic.

d. .33. Use the rule of thumb $|r| > \frac{2}{\sqrt{n}}$ to determine if the correlation is 'large'.

12.80 n = 47 with three independent variables. One of the independent variables has a correlation of .95 with the dependent variable.
Correlation between the independent variable and the dependent variable is not necessarily evidence of a small student's t statistic. A high correlation among the *independent* variables could result in a very small student's t statistic as the correlation creates a high variance.

12.82–12.84 Reports can be written by following the extended Case Study on the data file Cotton – see Section 12.9

12.86 **Regression Analysis: y_FemaleLFPR versus x1_income, x2_yrsedu, ...**

```
The regression equation is
y_FemaleLFPR = 0.2 +0.000406 x1_income + 4.84 x2_yrsedu - 1.55 x3_femaleun
Predictor          Coef      SE Coef         T         P        VIF
Constant           0.16        34.91      0.00     0.996
x1_incom      0.0004060    0.0001736      2.34     0.024        1.2
x2_yrsed          4.842        2.813      1.72     0.092        1.5
x3_femal        -1.5543       0.3399     -4.57     0.000        1.3

S = 3.048         R-Sq = 54.3%      R-Sq(adj) = 51.4%

Analysis of Variance
Source            DF          SS          MS         F          P
Regression         3      508.35      169.45     18.24      0.000
Residual Error    46      427.22        9.29
Total             49      935.57
```

12.88 **Regression Analysis: y_manufgrowt versus x1_aggrowth, x2_exportgro, ...**

```
The regression equation is
y_manufgrowth = 2.15 + 0.493 x1_aggrowth + 0.270 x2_exportgrowth
          - 0.117 x3_inflation
Predictor        Coef    SE Coef        T       P      VIF
Constant       2.1505     0.9695     2.22   0.032
x1_aggro       0.4934     0.2020     2.44   0.019      1.0
x2_expor      0.26991    0.06494     4.16   0.000      1.0
x3_infla     -0.11709    0.05204    -2.25   0.030      1.0

S = 3.624       R-Sq = 39.3%     R-Sq(adj) = 35.1%
Analysis of Variance
Source           DF        SS        MS       F       P
Regression        3    373.98    124.66    9.49   0.000
Residual Error   44    577.97     13.14
Total            47    951.95

Source       DF     Seq SS
x1_aggro      1      80.47
x2_expor      1     227.02
x3_infla      1      66.50
```

12.90 The analysis of variance table identifies how the total variability of the dependent variable (SST) is split up between the portion of variability that is explained by the regression model (SSR) and the part that is unexplained (SSE). The Coefficient of Determination (R^2) is derived as the ratio of SSR to SST. The analysis of variance table also computes the F statistic for the test of the significance of the overall regression – whether all of the slope coefficients are jointly equal to zero. The associated p-value is also generally reported in this table.

12.92 If one model contains more explanatory variables, then SST remains the same for both models but SSR will be higher for the model with more explanatory variables. Since SST = $SSR_1 + SSE_1$ which is equivalent to $SSR_2 + SSE_2$ and given that $SSR_2 > SSR_1$, then $SSE_1 > SSE_2$. Hence, the coefficient of determination will be higher with a greater number of explanatory variables and the coefficient of determination must be interpreted in conjunction with whether or not the regression slope coefficients on the explanatory variables are significantly different from zero.

12.94
$$\sum e_i = \sum (y_i - a - b_1 x_{1i} - b_2 x_{2i})$$
$$\sum e_i = \sum (y_i - \bar{y} + b_1 \bar{x}_{1i} + b_2 \bar{x}_{2i} - b_1 x_{1i} - b_2 x_{2i})$$
$$\sum e_i = n\bar{y} - n\bar{y} + nb_1\bar{x}_1 + nb_2\bar{x}_2 - nb_1\bar{x}_1 - nb_2\bar{x}_2$$
$$\sum e_i = 0$$

12.96 a. All else equal, an increase of one question results in a decrease of 1.834 in expected percentage of responses received. All else equal, an increase in one word in length of the questionnaire results in a decrease of .016 in expected percentage of responses received.

b. 63.7% of the variability in the percentage of responses received can be explained by the variability in the number of questions asked and the number of words

c. $H_0: \beta_1 = \beta_2 = 0, H_1:$ At least one $\beta_i \neq 0, (i = 1,2)$

$$F = \frac{n-k-1}{k}\frac{R^2}{1-R^2} = \frac{27}{2}\frac{.637}{1-.637} = 23.69$$

$F_{2,27,.01} = 5.49$, Therefore, reject H_0 at the 1% level

d. $t_{27,.005} = 2.771$, 99% CI: $-1.8345 \pm 2.771(.6349)$. –3.5938 up to -.0752

e. t = -1.78, $t_{27,.05/.025}$ = -1.703, -2.052.

Therefore, reject H_0 at the 5% level but not at the 2.5% level

12.98 **Regression Analysis: y_rating versus x1_expgrade, x2_Numstudents**

```
The regression equation is
y_rating = - 0.200 + 1.41 x1_expgrade - 0.0158 x2_Numstudents
Predictor        Coef       SE Coef          T        P        VIF
Constant      -0.2001        0.6968      -0.29    0.777
x1_expgr       1.4117        0.1780       7.93    0.000        1.5
x2_Numst    -0.015791      0.003783      -4.17    0.001        1.5

S = 0.1866        R-Sq = 91.5%      R-Sq(adj) = 90.5%
Analysis of Variance
Source            DF          SS         MS        F        P
Regression         2      6.3375     3.1687    90.99    0.000
Residual Error    17      0.5920     0.0348
Total             19      6.9295
```

12.100 a. All else equal, each extra point in the student's expected score leads to an expected increase of .469 in the actual score

b. $t_{103,.025}$=1.98, therefore, the 95% CI: $3.369 \pm 1.98(.456) = 2.4661$ up to 4.2719

c. $H_0: \beta_3 = 0, H_1: \beta_3 \neq 0$, $t = \frac{3.054}{1.457} = 2.096$, $t_{103,.025} = 1.96$

Therefore, reject H_0 at the 5% level

d. 68.6% of the variation in the exam scores is explained by their linear dependence on the student's expected score, hours per week spent working on the course and the student's grade point average

e. $H_0: \beta_1 = \beta_2 = \beta_3 = 0, H_1:$ At least one $\beta_i \neq 0, (i = 1,2,3)$

$$F = \frac{n-k-1}{k}\frac{R^2}{1-R^2} = \frac{103}{3}\frac{.686}{1-.686} = 75.008, F_{3,103,.01} = 3.95$$

Reject H_0 at any common levels of alpha

f. $R = \sqrt{.686} = .82825$

g. $\hat{Y} = 2.178 + .469(80) + 3.369(8) + 3.054(3) = 75.812$

12.102 a. $t_{2669,.05} = 1.645$, therefore, the 90% CI: $480.04 \pm 1.645(224.9) = 110.0795$ up to 850.0005

b. $t_{2669,.005} = 2.576$, therefore, the 99% CI: $1350.3 \pm 2.576(212.3) =$ 803.4152 up to 1897.1848

c. $H_0 : \beta_8 = 0, H_1 : \beta_8 < 0$, $t = \frac{-891.67}{180.87} = -4.9299$

$t_{2669,.005} = 2.576$, therefore, reject H_0 at the .5% level

d. $H_0 : \beta_9 = 0, H_1 : \beta_9 > 0$, $t = \frac{722.95}{110.98} = 6.5142$

$t_{2669,.005} = 2.576$, therefore, reject H_0 at the .5% level

e. 52.39% of the variability in minutes played in the season can be explained by the variability in all 9 variables.

f. $R = \sqrt{.5239} = .7238$

12.104 A report can be written by following the Case Study and testing the significance of the model. See section 12.9

12.106 **Correlations: Salary, age, Experien, yrs_asoc, yrs_full, Sex_1Fem, Market, C8**

	Salary	age	Experien	yrs_asoc	yrs_full	Sex_1Fem	Market
age	0.749						
	0.000						
Experien	0.883	0.877					
	0.000	0.000					
yrs_asoc	0.698	0.712	0.803				
	0.000	0.000	0.000				
yrs_full	0.777	0.583	0.674	0.312			
	0.000	0.000	0.000	0.000			
Sex_1Fem	-0.429	-0.234	-0.378	-0.367	-0.292		
	0.000	0.004	0.000	0.000	0.000		
Market	0.026	-0.134	-0.150	-0.113	-0.017	0.062	
	0.750	0.103	0.067	0.169	0.833	0.453	
C8	-0.029	-0.189	-0.117	-0.073	-0.043	-0.094	-0.107
	0.721	0.020	0.155	0.373	0.598	0.254	0.192

The correlation matrix indicates that several of the independent variables are likely to be significant, however, multicollinearity is also a likely result. The regression model with all independent variables is:

Regression Analysis: Salary versus age, Experien, ...

```
The regression equation is
Salary = 23725 - 40.3 age + 357 Experien + 263 yrs_asoc +
493 yrs_full
           - 954 Sex_1Fem + 3427 Market + 1188 C8
Predictor         Coef      SE Coef          T        P         VIF
Constant         23725         1524      15.57    0.000
age             -40.29        44.98      -0.90    0.372         4.7
Experien        356.83        63.48       5.62    0.000        10.0
yrs_asoc        262.50        75.11       3.49    0.001         4.0
yrs_full        492.91        59.27       8.32    0.000         2.6
Sex_1Fem        -954.1        487.3      -1.96    0.052         1.3
Market          3427.2        754.1       4.54    0.000         1.1
C8              1188.4        597.5       1.99    0.049         1.1

S = 2332          R-Sq = 88.2%      R-Sq(adj) = 87.6%
Analysis of Variance
Source            DF           SS          MS         F          P
Regression         7   5776063882   825151983    151.74      0.000
Residual Error   142    772162801     5437766
Total            149   6548226683

Source        DF      Seq SS
age            1  3669210599
Experien       1  1459475287
yrs_asoc       1     1979334
yrs_full       1   500316356
Sex_1Fem       1    22707368
Market         1   100860164
```

Since age is insignificant and has the smallest t-statistics, it is removed from the model.

The conditional F test for age is:

$$F_{X_2} = \frac{SSR_F - SSR_R}{s^2_{Y|X}} = \frac{5,766,064,000 - 5,771,700,736}{(2332)^2} = .80$$

Which is well below any common critical value of F. Thus, age is removed from the model. The remaining independent variables are all significant at the .05 level of significance and hence, become the final regression model. Residual analysis to determine if the assumption of linearity holds true follows:

Regression Analysis: Salary versus Experien, yrs_asoc, ...

```
The regression equation is
Salary = 22455 + 324 Experien + 258 yrs_asoc + 491 yrs_full - 1043 Sex_1Fem
           + 3449 Market + 1274 C8
Predictor         Coef     SE Coef          T        P         VIF
Constant       22455.2       557.7      40.26    0.000
Experien        324.24       51.99       6.24    0.000         6.7
yrs_asoc        257.88       74.88       3.44    0.001         4.0
yrs_full        490.97       59.19       8.29    0.000         2.6
Sex_1Fem       -1043.4       476.7      -2.19    0.030         1.2
Market          3449.4       753.2       4.58    0.000         1.1
C8              1274.5       589.3       2.16    0.032         1.1

S = 2330          R-Sq = 88.1%      R-Sq(adj) = 87.6%
Analysis of Variance
Source            DF           SS          MS         F          P
Regression         6   5771700580   961950097    177.15      0.000
Residual Error   143    776526103     5430252
Total            149   6548226683
```

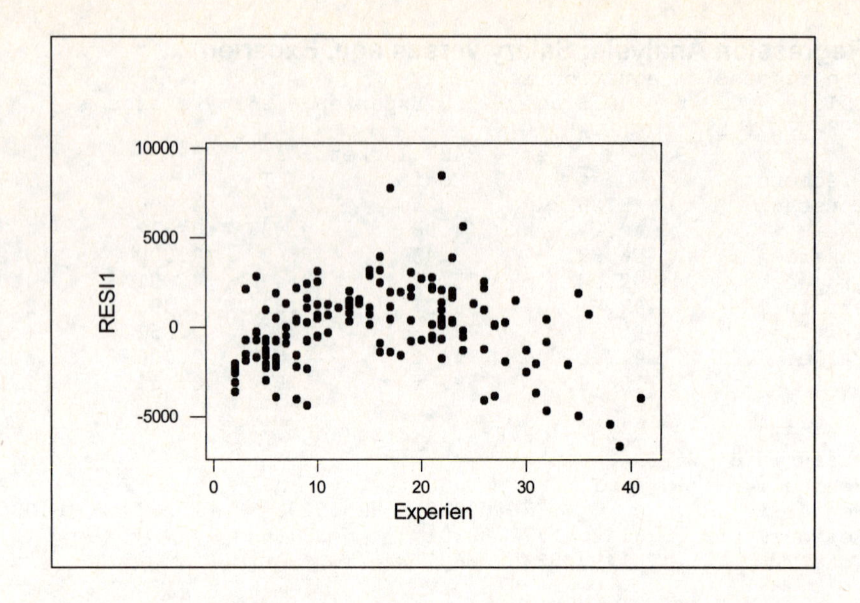
10000
5000
0
-5000
RESI1
0
10
20
30
40
Experien

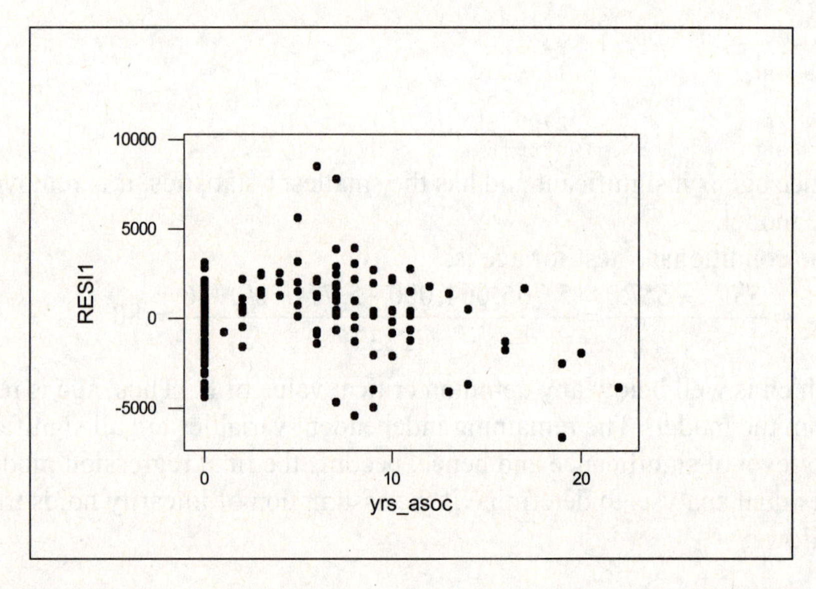
10000
5000
0
-5000
RESI1
0
10
20
yrs_asoc

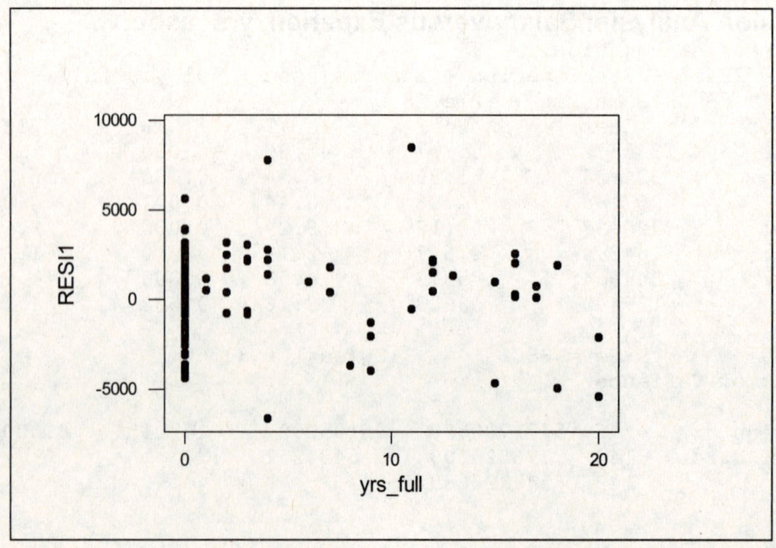
10000
5000
0
-5000
RESI1
0
10
20
yrs_full

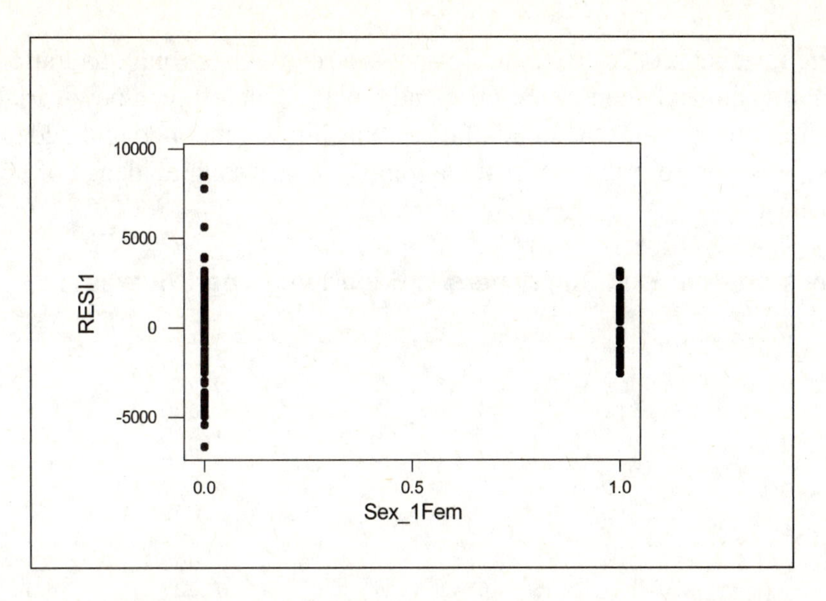
10000
5000
0
-5000
RESI1
0.0
0.5
1.0
Sex_1Fem

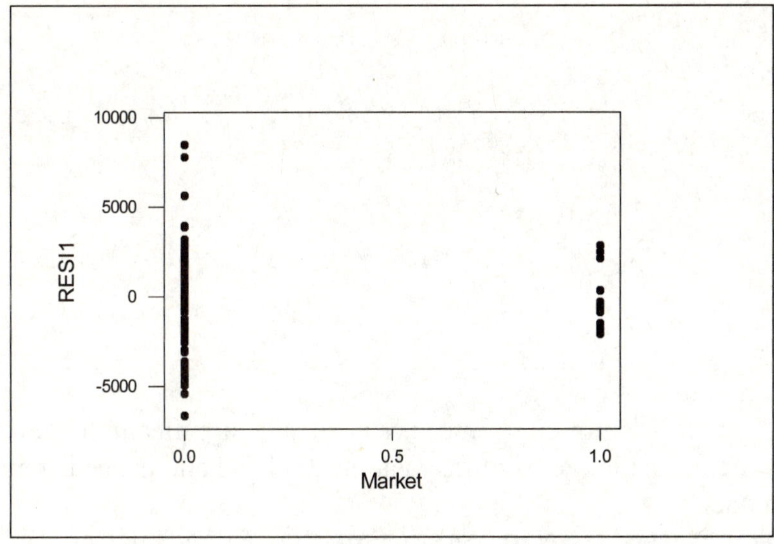
10000
5000
0
-5000
RESI1
0.0
0.5
1.0
Market

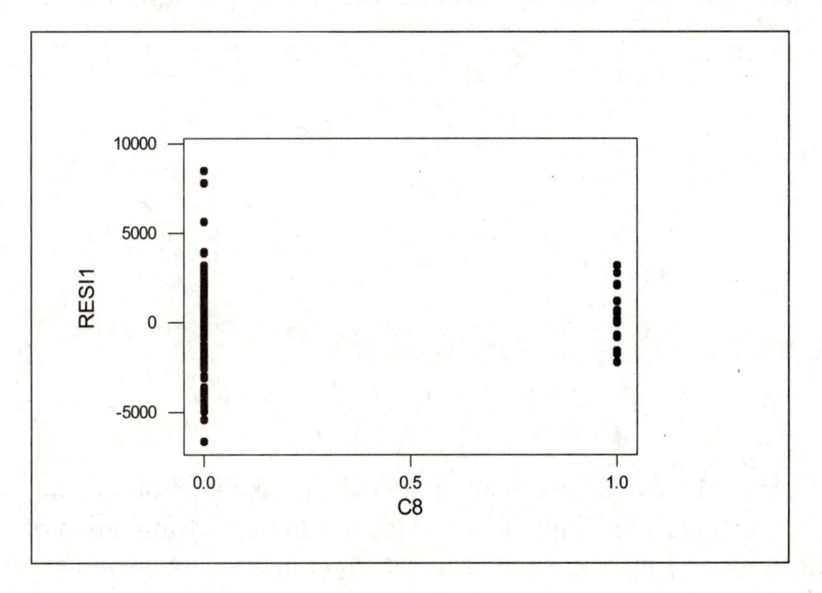
10000
5000
0
-5000
RESI1
0.0
0.5
1.0
C8

The residual plot for Experience shows a relatively strong quadratic relationship between Experience and Salary. Therefore, a new variable, taking into account the quadratic relationship is generated and added to the model. None of the other residual plots shows strong evidence of non-linearity.

Regression Analysis: Salary versus Experien, ExperSquared, ...

```
The regression equation is
Salary = 18915 + 875 Experien - 15.9 ExperSquared + 222 yrs_asoc + 612 yrs_full
         - 650 Sex_1Fem + 3978 Market + 1042 C8
Predictor        Coef     SE Coef          T        P         VIF
Constant      18915.2       583.2      32.43    0.000
Experien       875.35       72.20      12.12    0.000        20.6
ExperSqu      -15.947       1.717      -9.29    0.000        16.2
yrs_asoc       221.58       59.40       3.73    0.000         4.0
yrs_full       612.10       48.63      12.59    0.000         2.8
Sex_1Fem       -650.1       379.6      -1.71    0.089         1.2
Market         3978.3       598.8       6.64    0.000         1.1
C8             1042.3       467.1       2.23    0.027         1.1

S = 1844        R-Sq = 92.6%     R-Sq(adj) = 92.3%
Analysis of Variance
Source            DF          SS         MS          F          P
Regression         7  6065189270  866455610     254.71      0.000
Residual Error   142   483037413    3401672
Total            149  6548226683

Source         DF      Seq SS
Experien        1  5109486518
ExperSqu        1    91663414
yrs_asoc        1    15948822
yrs_full        1   678958872
Sex_1Fem        1    12652358
Market          1   139540652
C8              1    16938635
```

The squared term for experience is statistically significant; however, the Sex_1Fem is no longer significant at the .05 level and hence is removed from the model:

Regression Analysis: Salary versus Experien, ExperSquared, ...

```
The regression equation is
Salary = 18538 + 888 Experien - 16.3 ExperSquared + 237 yrs_asoc + 624 yrs_full
         + 3982 Market + 1145 C8
Predictor        Coef     SE Coef          T        P         VIF
Constant      18537.8       543.6      34.10    0.000
Experien       887.85       72.32      12.28    0.000        20.4
ExperSqu      -16.275       1.718      -9.48    0.000        16.0
yrs_asoc       236.89       59.11       4.01    0.000         3.9
yrs_full       624.49       48.41      12.90    0.000         2.8
Market         3981.8       602.9       6.60    0.000         1.1
C8             1145.4       466.3       2.46    0.015         1.0

S = 1857        R-Sq = 92.5%     R-Sq(adj) = 92.2%
Analysis of Variance
Source            DF          SS          MS          F          P
Regression         6  6055213011  1009202168     292.72      0.000
Residual Error   143   493013673     3447648
Total            149  6548226683
```

This is the final model with all of the independent variables being conditionally significant, including the quadratic transformation of Experience. This would indicate that a non-linear relationship exists between experience and salary.

12.108 a. Correlation matrix:

Correlations: deaths, vehwt, impcars, lghttrks, carage

```
           deaths     vehwt  impcars lghttrks
vehwt       0.244
            0.091

impcars    -0.284    -0.943
            0.048     0.000

lghttrks    0.726     0.157   -0.175
            0.000     0.282    0.228

carage     -0.422     0.123    0.011   -0.329
            0.003     0.400    0.943    0.021
```

Crash deaths are positively related to vehicle weight and percentage of light trucks and negatively related to percent imported cars and car age. Light trucks will have the strongest linear association of any independent variable followed by car age. Multicollinearity is likely to exist due to the strong correlation between impcars and vehicle weight.

b. **Regression Analysis: deaths versus vehwt, impcars, lghttrks, carage**

```
The regression equation is
deaths = 2.60 +0.000064 vehwt - 0.00121 impcars + 0.00833 lghttrks
            - 0.0395 carage
Predictor          Coef    SE Coef        T       P       VIF
Constant          2.597      1.247     2.08   0.043
vehwt         0.0000643  0.0001908     0.34   0.738      10.9
impcars       -0.001213   0.005249    -0.23   0.818      10.6
lghttrks       0.008332   0.001397     5.96   0.000       1.2
carage         -0.03946    0.01916    -2.06   0.045       1.4

S = 0.05334      R-Sq = 59.5%      R-Sq(adj) = 55.8%
Analysis of Variance
Source           DF         SS          MS        F       P
Regression        4   0.183634    0.045909    16.14   0.000
Residual Error   44   0.125174    0.002845
Total            48   0.308809
```

Light trucks is a significant positive variable. Since impcars has the smallest t-statistic, it is removed from the model:

Regression Analysis: deaths versus vehwt, lghttrks, carage

```
The regression equation is
deaths = 2.55 +0.000106 vehwt + 0.00831 lghttrks - 0.0411 carage
Predictor          Coef     SE Coef        T       P       VIF
Constant          2.555       1.220     2.09   0.042
vehwt        0.00010622  0.00005901     1.80   0.079       1.1
lghttrks       0.008312    0.001380     6.02   0.000       1.2
carage         -0.04114     0.01754    -2.34   0.024       1.2

S = 0.05277      R-Sq = 59.4%      R-Sq(adj) = 56.7%

Analysis of Variance
Source           DF         SS          MS        F       P
Regression        3   0.183482    0.061161    21.96   0.000
Residual Error   45   0.125326    0.002785
Total            48   0.308809
```

Also, remove vehicle weight using the same argument:

Regression Analysis: deaths versus lghttrks, carage

The regression equation is

deaths = 2.51 + 0.00883 lghttrks - 0.0352 carage

Predictor	Coef	SE Coef	T	P	VIF
Constant	2.506	1.249	2.01	0.051	
lghttrks	0.008835	0.001382	6.39	0.000	1.1
carage	-0.03522	0.01765	-2.00	0.052	1.1

S = 0.05404 R-Sq = 56.5% R-Sq(adj) = 54.6%

Analysis of Variance

Source	DF	SS	MS	F	P
Regression	2	0.174458	0.087229	29.87	0.000
Residual Error	46	0.134351	0.002921		
Total	48	0.308809			

The model has light trucks and car age as the significant variables. Note that car age is marginally significant (p-value of .052) and hence could also be dropped from the model.

c. The regression modeling indicates that the percentage of light trucks is conditionally significant in all of the models and hence is an important predictor in the model. Car age and imported cars are marginally significant predictors when only light trucks is included in the model.

12.110 a. Correlation matrix and descriptive statistics

Correlations: hseval, sizehse, Taxhse, Comper, incom72, totexp

	hseval	sizehse	Taxhse	Comper	incom72
sizehse	0.542				
	0.000				
Taxhse	0.248	0.289			
	0.019	0.006			
Comper	-0.335	-0.278	-0.114		
	0.001	0.008	0.285		
incom72	0.426	0.393	0.261	-0.198	
	0.000	0.000	0.013	0.062	
totexp	0.261	-0.022	0.228	0.269	0.376
	0.013	0.834	0.030	0.010	0.000

The correlation matrix shows that multicollinearity is not likely to be a problem in this model since all of the correlations among the independent variables are relatively low.

Descriptive Statistics: hseval, sizehse, Taxhse, Comper, incom72, totexp

Variable	N	Mean	Median	TrMean	StDev	SE Mean
hseval	90	21.031	20.301	20.687	4.957	0.522
sizehse	90	5.4778	5.4000	5.4638	0.2407	0.0254
Taxhse	90	130.13	131.67	128.31	48.89	5.15
Comper	90	0.16211	0.15930	0.16206	0.06333	0.00668
incom72	90	3360.9	3283.0	3353.2	317.0	33.4
totexp	90	1488848	1089110	1295444	1265564	133402

Variable	Minimum	Maximum	Q1	Q3
hseval	13.300	35.976	17.665	24.046
sizehse	5.0000	6.2000	5.3000	5.6000
Taxhse	35.04	399.60	98.85	155.19
Comper	0.02805	0.28427	0.11388	0.20826
incom72	2739.0	4193.0	3114.3	3585.3
totexp	361290	7062330	808771	1570275

The range for applying the regression model (variable means + / - 2 standard errors):

Hseval	21.03 +/- 2(4.957) = 11.11 to 30.94
Sizehse	5.48 +/- 2(.24) = 5.0 to 5.96
Taxhse	130.13 +/- 2(48.89) = 32.35 to 227.91
Comper	.16 +/- 2(.063) = .034 to .286
Incom72	3361 +/- 2(317) = 2727 to 3995
Totexp	1488848 +/- 2(1265564) = not a good approximation

b. Regression models:

Regression Analysis: hseval versus sizehse, Taxhse, ...

```
The regression equation is
hseval = - 31.1 + 9.10 sizehse - 0.00058 Taxhse - 22.2 Comper + 0.00120 incom72
         +0.000001 totexp
Predictor          Coef      SE Coef         T        P        VIF
Constant         -31.07        10.09     -3.08    0.003
sizehse           9.105        1.927      4.72    0.000        1.3
Taxhse        -0.000584     0.008910     -0.07    0.948        1.2
Comper          -22.197        7.108     -3.12    0.002        1.3
incom72        0.001200     0.001566      0.77    0.445        1.5
totexp       0.00000125   0.00000038      3.28    0.002        1.5

S = 3.785       R-Sq = 45.0%       R-Sq(adj) = 41.7%

Analysis of Variance
Source            DF          SS         MS        F        P
Regression         5      982.98     196.60    13.72    0.000
Residual Error    84     1203.65      14.33
Total             89     2186.63
```

Taxhse is not conditionally significant, nor is income; however, dropping one variable at a time, eliminate Taxhse first, then eliminate income.

Regression Analysis: hseval versus sizehse, Comper, totexp

```
The regression equation is
hseval = - 29.9 + 9.61 sizehse - 23.5 Comper +0.000001 totexp
Predictor          Coef      SE Coef         T        P        VIF
Constant        -29.875        9.791     -3.05    0.003
sizehse           9.613        1.724      5.58    0.000        1.1
Comper          -23.482        6.801     -3.45    0.001        1.2
totexp       0.00000138   0.00000033      4.22    0.000        1.1

S = 3.754       R-Sq = 44.6%       R-Sq(adj) = 42.6%

Analysis of Variance
Source            DF          SS         MS        F        P
Regression         3      974.55     324.85    23.05    0.000
Residual Error    86     1212.08      14.09
Total             89     2186.63
```

This is the final regression model. All of the independent variables are conditionally significant.

Both the size of house and total government expenditures enhances market value of homes while the percent of commercial property tends to reduce market values of homes.

c. In the final regression model, the tax variable was not found to be conditionally significant and hence it is difficult to support the developer's claim.

12.112 a. **Correlations: Res. Invest., Prime Rate, Federal Funds Rate, GDP, Money Supply, Govt. Spending**

	Res. Inv	Prime Ra	Federal	GDP	Money Su
Prime Ra	0.213				
	0.001				
Federal	0.080	0.960			
	0.220	0.000			
GDP	0.957	0.333	0.170		
	0.000	0.000	0.009		
Money Su	0.944	0.246	0.073	0.987	
	0.000	0.000	0.262	0.000	
Govt. Sp	0.935	0.361	0.209	0.986	0.967
	0.000	0.000	0.001	0.000	0.000

The correlation matrix shows that both interest rates have a significant positive impact on residential investment. The money supply, GDP and government spending also have a significant linear association with residential investment. Note the high correlation between the two interest rate variables, which, as expected, would create significant problems if both variables are included in the regression model. Hence, the interest rates will be developed in two separate models.

Regression Analysis: Res. Invest. versus GDP, Prime Rate, Money Supply, Govt. Spending

```
The regression equation is
Res. Invest. = 50.3 + 0.0789 GDP - 5.90 Prime Rate - 0.0340
Money Supply - 0.0792 Govt. Spending
```

Predictor	Coef	StDev	T	P
Constant	50.29	15.25	3.30	0.001
GDP	0.078923	0.007394	10.67	0.000
Prime Ra	-5.9020	0.7707	-7.66	0.000
Money Su	-0.033979	0.007378	-4.61	0.000
Govt. Sp	-0.07918	0.03009	-2.63	0.009

S = 31.90   R-Sq = 93.6%   R-Sq(adj) = 93.5%

Analysis of Variance

Source	DF	SS	MS	F	P
Regression	4	3418627	854657	839.93	0.000
Residual Error	231	235050	1018		
Total	235	3653676			

This will be the final model with prime rate as the interest rate variable since all of the independent variables are conditionally significant. Note the significant multicollinearity that exists between the independent variables.

Regression Analysis Res. Invest. versus GDP, Federal Funds Rate, Money Supply, Govt. Spending

```
The regression equation is
Res. Invest. = 45.1 - 4.23 Federal Funds Rate + 0.0740 GDP
         - 0.0295 Money Supply - 0.0766 Govt. Spending

Predictor          Coef        StDev         T        P
Constant          45.07        16.90      2.67    0.008
Federal         -4.2287       0.7815     -5.41    0.000
GDP            0.073987     0.007793      9.49    0.000
Money Su      -0.029509     0.008067     -3.66    0.000
Govt. Sp       -0.07662      0.03195     -2.40    0.017

S = 33.65        R-Sq = 92.8%      R-Sq(adj) = 92.7%

Analysis of Variance

Source            DF          SS          MS         F        P
Regression         4     3392102      848025    748.90    0.000
Residual Error   231      261575        1132
Total            235     3653676
```

The model with the federal funds rate as the interest rate variable is also the final model with all of the independent variables conditionally significant. Again, high correlation among the independent variables will be a problem with this regression model.

b. 95% confidence intervals for the slope coefficients on the interest rate term:

Prime interest rate as the interest rate variable:

$b_1 \pm t_{n-K-1,\alpha/2} s_{b_1}$: -5.9020 +/- 1.97(.7707) = -5.9020 +/- 1.518279 or (-7.420, -4.384)

Federal funds rate as the interest rate variable:

$b_1 \pm t_{n-K-1,\alpha/2} s_{b_1}$: -4.2287 +/- 1.97(.7815) = -4.2287 +/- 1.539555 or (-5.768, -2.689)

12.114 a. **Correlations: Salary, age, yrs_asoc, yrs_full, Sex_1Fem, Market, C8**

```
            Salary      age yrs_asoc yrs_full Sex_1Fem   Market
age          0.749
             0.000
yrs_asoc     0.698    0.712
             0.000    0.000
yrs_full     0.777    0.583    0.312
             0.000    0.000    0.000
Sex_1Fem    -0.429   -0.234   -0.367   -0.292
             0.000    0.004    0.000    0.000
Market       0.026   -0.134   -0.113   -0.017    0.062
             0.750    0.103    0.169    0.833    0.453
C8          -0.029   -0.189   -0.073   -0.043   -0.094   -0.107
             0.721    0.020    0.373    0.598    0.254    0.192
```

The correlation matrix indicates several independent variables that should provide good explanatory power in the regression model. We would expect that age, years at Associate professor and years at full professor are likely to be conditionally significant:

Regression Analysis: Salary versus age, yrs_asoc, ...

```
The regression equation is
Salary = 21107 + 105 age + 532 yrs_asoc + 690 yrs_full - 1312 Sex_1Fem
           + 2854 Market + 1101 C8
Predictor          Coef      SE Coef          T        P           VIF
Constant          21107         1599      13.20    0.000
age              104.59        40.62       2.58    0.011           3.1
yrs_asoc         532.27        63.66       8.36    0.000           2.4
yrs_full         689.93        52.66      13.10    0.000           1.7
Sex_1Fem        -1311.8        532.3      -2.46    0.015           1.3
Market           2853.9        823.3       3.47    0.001           1.0
C8               1101.0        658.1       1.67    0.097           1.1

S = 2569        R-Sq = 85.6%     R-Sq(adj) = 85.0%

Analysis of Variance
Source            DF           SS          MS          F           P
Regression         6   5604244075   934040679     141.49       0.000
Residual Error   143    943982608     6601277
Total            149   6548226683
```

Dropping the C8 variable yields:

Regression Analysis: Salary versus age, yrs_asoc, ...

```
The regression equation is
Salary = 21887 + 90.0 age + 539 yrs_asoc + 697 yrs_full - 1397 Sex_1Fem
           + 2662 Market
Predictor          Coef      SE Coef          T        P           VIF
Constant          21887         1539      14.22    0.000
age               90.02        39.92       2.26    0.026           3.0
yrs_asoc         539.48        63.91       8.44    0.000           2.4
yrs_full         697.35        52.80      13.21    0.000           1.7
Sex_1Fem        -1397.2        533.2      -2.62    0.010           1.2
Market           2662.3        820.3       3.25    0.001           1.0

S = 2585        R-Sq = 85.3%     R-Sq(adj) = 84.8%

Analysis of Variance
Source            DF           SS          MS          F           P
Regression         5   5585766862  1117153372     167.14       0.000
Residual Error   144    962459821     6683749
Total            149   6548226683
```

This is the final model. All of the independent variables are conditionally significant and the model explains a sizeable portion of the variability in salary.

b. To test the hypothesis that the rate of change in female salaries as a function of age is less than the rate of change in male salaries as a function of age, the dummy variable Sex_1Fem is used to see if the slope coefficient for age (X1) is different for males and females. The following model is used:

$$Y = \beta_0 + (\beta_1 + \beta_6 X_4)X_1 + \beta_2 X_2 + \beta_3 X_3 + \beta_4 X_4 + \beta_5 X_5$$
$$= \beta_0 + \beta_1 X_1 + \beta_6 X_4 X_1 + \beta_2 X_2 + \beta_3 X_3 + \beta_4 X_4 + \beta_5 X_5$$

Create the variable X_4X_1 and then test for conditional significance in the regression model. If it proves to be a significant predictor of salaries then there is strong evidence to conclude that the rate of change in female salaries as a function of age is different than for males:

Regression Analysis: Salary versus age, femage, ...

```
The regression equation is
Salary = 22082 + 85.1 age + 11.7 femage + 543 yrs_asoc + 701 yrs_full
         - 1878 Sex_1Fem + 2673 Market
Predictor         Coef      SE Coef          T          P          VIF
Constant         22082         1877      11.77      0.000
age              85.07        48.36       1.76      0.081          4.4
femage           11.66        63.89       0.18      0.855         32.2
yrs_asoc        542.85        66.73       8.13      0.000          2.6
yrs_full        701.35        57.35      12.23      0.000          2.0
Sex_1Fem         -1878         2687      -0.70      0.486         31.5
Market          2672.8        825.1       3.24      0.001          1.0

S = 2594        R-Sq = 85.3%       R-Sq(adj) = 84.7%
Analysis of Variance
Source            DF          SS           MS          F          P
Regression         6  5585990999    930998500     138.36      0.000
Residual Error   143   962235684      6728921
Total            149  6548226683
```

The regression shows that the newly created variable of femage is not conditionally significant. Thus, we cannot conclude that the rate of change in female salaries as a function of age differs from that of male salaries.

12.116 a. Correlation matrix:

Correlations: EconGPA, sex, Acteng, ACTmath, ACTss, ACTcomp, HSPct

```
           EconGPA       sex    Acteng   ACTmath     ACTss  ACTcomp
sex          0.187
             0.049
Acteng       0.387     0.270
             0.001     0.021
ACTmath      0.338    -0.170     0.368
             0.003     0.151     0.001
ACTss        0.442    -0.105     0.448     0.439
             0.000     0.375     0.000     0.000
ACTcomp      0.474    -0.084     0.650     0.765     0.812
             0.000     0.478     0.000     0.000     0.000
HSPct        0.362     0.216     0.173     0.290     0.224     0.230
             0.000     0.026     0.150     0.014     0.060     0.053
```

There exists a positive relationship between EconGPA and all of the independent variables, which is expected. Note that there is a high correlation between the composite ACT score and the individual components, which is again, as expected. Thus, high correlation among the independent variables is likely to be a serious concern in this regression model.

Regression Analysis: EconGPA versus sex, Acteng, ...

```
The regression equation is
EconGPA = - 0.050 + 0.261 sex + 0.0099 Acteng + 0.0064 ACTmath +
0.0270 ACTss
          + 0.0419 ACTcomp + 0.00898 HSPct
71 cases used 41 cases contain missing values
Predictor        Coef      SE Coef        T         P        VIF
Constant      -0.0504       0.6554    -0.08     0.939
sex            0.2611       0.1607     1.62     0.109        1.5
Acteng        0.00991      0.02986     0.33     0.741        2.5
ACTmath       0.00643      0.03041     0.21     0.833        4.3
ACTss         0.02696      0.02794     0.96     0.338        4.7
ACTcomp       0.04188      0.07200     0.58     0.563       12.8
HSPct        0.008978     0.005716     1.57     0.121        1.4

S = 0.4971       R-Sq = 34.1%      R-Sq(adj) = 27.9%
Analysis of Variance
Source            DF          SS          MS         F        P
Regression         6      8.1778      1.3630      5.52    0.000
Residual Error    64     15.8166      0.2471
Total             70     23.9945
```

As expected, high correlation among the independent variables is affecting the results. A strategy of dropping the variable with the lowest t-statistic with each successive model causes the dropping of the following variables (in order): 1) ACTmath, 2) ACTeng, 3) ACTss, 4) HSPct. The two variables that remain are the final model of gender and ACTcomp:

Regression Analysis: EconGPA versus sex, ACTcomp

```
The regression equation is
EconGPA = 0.322 + 0.335 sex + 0.0978 ACTcomp
73 cases used 39 cases contain missing values
Predictor        Coef      SE Coef        T         P        VIF
Constant       0.3216       0.5201     0.62     0.538
sex            0.3350       0.1279     2.62     0.011        1.0
ACTcomp       0.09782      0.01989     4.92     0.000        1.0

S = 0.4931       R-Sq = 29.4%      R-Sq(adj) = 27.3%
Analysis of Variance
Source            DF          SS          MS         F        P
Regression         2      7.0705      3.5352     14.54    0.000
Residual Error    70     17.0192      0.2431
Total             72     24.0897
```

Both independent variables are conditionally significant.

b. The model could be used in college admission decisions by creating a predicted GPA in economics based on sex and ACT comp scores. This predicted GPA could then be used with other factors in deciding admission. Note that this model predicts that females will outperform males with equal test scores. Using this model as the only source of information may lead to charges of unequal treatment.

Chapter 13:

Additional Topics in Regression Analysis

13.2 $Y_i = \beta_0 + \beta_1 X_{1i} + \beta_2 X_{2i} + \beta_3 X_{3i} + \beta_4 X_{4i} + \beta_5 X_{5i} + \varepsilon_i$

where Y_i = wages

X_1 = Years of experience

X_2 = 1 for Germany, 0 otherwise

X_3 = 1 for Great Britain, 0 otherwise

X_4 = 1 for Japan, 0 otherwise

X_5 = 1 for Turkey, 0 otherwise

The excluded category consists of wages in the United States

13.4 a. For any observation, the values of the dummy variables sum to one. Since the equation has an intercept term, there is perfect multicollinearity and the existence of the "dummy variable trap".

b. β_3 measures the expected difference between demand in the first and fourth quarters, all else equal. β_4 measures the expected difference between demand in the second and fourth quarters, all else equal. β_5 measures the expected difference between demand in the third and fourth quarters, all else equal.

13.6 $Y_i = \beta_0 + \beta_1 X_{1i} + \beta_2 X_{2i} + \beta_3 X_{3i} + \beta_4 X_{4i} + \beta_5 X_{5i} + \varepsilon_i$

where Y_i = per capita cereal sales

X_1 = cereal price

X_2 = price of competing cereals

X_3 = mean per capita income

X_4 = % college graduates

X_5 = mean annual temperature

X_6 = mean annual rainfall

X_7 = 1 for cities east of the Mississippi, 0 otherwise

X_8 = 1 for high per capita income, 0 otherwise

X_9 = 1 for intermediate per capita income, 0 otherwise

X_{10} = 1 for northwest, 0 otherwise

X_{11} = 1 for southwest, 0 otherwise

X_{12} = 1 for northeast, 0 otherwise

X_{13} = X_1X_7–interaction term between price and cities east of the Mississippi

The model specification includes continuous independent variables, dichotomous indicator variables and slope dummy variables. Based on economic demand theory, we would expect the coefficient on cereal price to be negative due to the law of demand. Prices of substitutes are expected to have a positive impact on per capita cereal sales. If the cereal is deemed a normal good, mean per capita income will have a positive impact on sales. The signs and sizes of other variables may be empirically determined. While the functional form can be linear, non-linearity could be introduced based on an initial analysis of the scatterplots of the relationships. High correlation among the independent variables could also be detected, for example, per capita income and % college graduates may very well be collinear. Several iterations of the model could be conducted to find the optimal combinations of variables.

13.8 Define the following variables for the experiment

Y = worker compensation

X_1 = years of experience

X_2 = job classification level

1. Apprentice
2. Professional
3. Master

X_3 = individual ability

X_4 = gender

1. male
2. female

X_5 = race

1. White
2. Black
3. Latino

Two different dependent variables can be developed from the salary data. Base compensation will be one analysis that can be conducted. The incremental salaries can also be analyzed. Dummy variables are required to analyze the impact of job classifications on salary. Discrimination can be measured by the size of the dummy variable on gender and on race. For each dummy variable, (k-1) categories are required to avoid the 'dummy variable trap.' The F-test for the significance of the overall regression will be utilized to determine whether the model has significant explanatory power. And the t-test for the significance of the individual regression slope coefficients will be utilized to determine the impact of each independent variable. Model diagnostics will be based on R-square and the behavior of the residuals.

13.10 What is the long term effect of a one unit increase in x in period t?

a. $\frac{\beta_j}{(1-\gamma)} = \frac{2}{(1-.34)} = 3.03$

b. $\frac{\beta_j}{(1-\gamma)} = \frac{2.5}{(1-.24)} = 3.289$

c. $\frac{\beta_j}{(1-\gamma)} = \frac{2}{(1-.64)} = 5.556$

d. $\frac{\beta_j}{(1-\gamma)} = \frac{4.3}{(1-.34)} = 6.515$

13.12 **Regression Analysis: Y Retail Sales versus X Income, Ylag1**

```
The regression equation is
Y  Retail Sales = 1752 + 0.367 X  Income + 0.053 Ylag1
21 cases used 1 cases contain missing values

Predictor         Coef     SE Coef          T        P
Constant        1751.6       500.0       3.50    0.003
X  Incom       0.36734     0.08054       4.56    0.000
Ylag1           0.0533      0.2035       0.26    0.796

S = 153.4        R-Sq = 91.7%      R-Sq(adj) = 90.7%
```

$t = \frac{.0533}{.2035} = .2619$; $t_{18,.10} = 1.33$, therefore, do not reject H_0 at the 20% level

13.14 **Regression Analysis: Y_%stocks versus X_Return, Y_lag%stocks**

```
The regression equation is
Y_%stocks = 1.65 + 0.228 X_Return + 0.950 Y_lag%stocks
24 cases used 1 cases contain missing values

Predictor         Coef     SE Coef          T        P
Constant         1.646       2.414       0.68    0.503
X_Return       0.22776     0.03015       7.55    0.000
Y_lag%st       0.94999     0.04306      22.06    0.000

S = 2.351        R-Sq = 95.9%      R-Sq(adj) = 95.5%
Analysis of Variance
Source              DF          SS          MS         F        P
Regression           2      2689.6      1344.8    243.38    0.000
Residual Error      21       116.0         5.5
Total               23      2805.6

Source        DF       Seq SS
X_Return       1          0.7
Y_lag%st       1       2688.9

Unusual Observations
Obs   X_Return   Y_%stock          Fit        SE Fit      Residual      St Resid
 20      -26.5     56.000       60.210         1.160        -4.210        -2.06R
```

13.16

Regression Analysis: Y_Birth versus X_1stmarriage, Y_lagBirth

```
The regression equation is
Y_Birth = 21262 + 0.485 X_1stmarriage + 0.192 Y_lagBirth

19 cases used 1 cases contain missing values

Predictor        Coef    SE Coef        T       P
Constant        21262      5720      3.72   0.002
X_1stmar       0.4854    0.1230      3.94   0.001
Y_lagBir       0.1923    0.1898      1.01   0.326

S = 2513         R-Sq = 93.7%     R-Sq(adj) = 93.0%
Analysis of Variance
Source           DF          SS          MS        F       P
Regression        2  1515082551   757541276   119.93   0.000
Residual Error   16   101062160     6316385
Total            18  1616144711

Source        DF      Seq SS
X_1stmar       1  1508597348
Y_lagBir       1     6485203

Unusual Observations
Obs   X_1stmar    Y_Birth        Fit      SE Fit    Residual    St Resid
 15     105235      95418      89340         982        6078       2.63R
```

13.18

Regression Analysis: Y_logCons versus X_LogDI, Y_laglogCons

```
The regression equation is
Y_logCons = 0.405 + 0.373 X_LogDI + 0.558 Y_laglogCons

28 cases used 1 cases contain missing values

Predictor        Coef    SE Coef        T       P
Constant       0.4049    0.1051      3.85   0.001
X_LogDI        0.3734    0.1075      3.47   0.002
Y_laglog       0.5577    0.1243      4.49   0.000

S = 0.03023      R-Sq = 99.6%     R-Sq(adj) = 99.6%
Analysis of Variance
Source           DF          SS          MS        F       P
Regression        2      6.1960      3.0980  3389.90   0.000
Residual Error   25      0.0228      0.0009
Total            27      6.2189

Source        DF      Seq SS
X_LogDI        1      6.1776
Y_laglog       1      0.0184

Unusual Observations
Obs    X_LogDI   Y_logCon        Fit      SE Fit    Residual    St Resid
  9       5.84    5.80814    5.72298     0.01074     0.08517       3.01R

Durbin-Watson statistic = 1.63
```

13.20 a. In the special case where the sample correlations between x_1 and x_2 is zero, the estimate for β_1 will be the same whether or not x_2 is included in the regression equation. In the simple linear regression of y on x_1, the intercept term will embody the influence of x_2 on y, under these special circumstances.

b. $$b_1 = \frac{\sum(x_{2i}-\bar{x}_2)^2\sum(x_{1i}-\bar{x}_1)(y_{1i}-\bar{y}) - \sum(x_{1i}-\bar{x}_1)(x_{2i}-\bar{x}_2)\sum(x_{2i}-\bar{x}_2)(y_i-\bar{y})}{\sum(x_{1i}-\bar{x}_1)^2\sum(x_{2i}-\bar{x}_2)^2 - [\sum(x_{1i}-\bar{x}_1)\sum(x_{2i}-\bar{x}_2)]^2}$$

If the sample correlation between x_1 and x_2 is zero, then $\sum(x_{1i}-\bar{x}_1)(x_{2i}-\bar{x}_2)=0$ and the slope coefficient equation can be simplified. The result is

$$b_1 = \frac{\sum(x_{1i}-\bar{x}_1)(y_{1i}-\bar{y})}{\sum(x_{1i}-\bar{x}_1)^2}$$ which is the estimated slope coefficient for the bivariate linear regression of y on x_1.

13.22

Results for: CITYDAT.XLS

Regression Analysis: hseval versus Comper, Homper, ...

```
The regression equation is
hseval = - 19.0 - 26.4 Comper - 12.1 Homper - 15.5 Indper + 7.22 sizehse
           + 0.00408 incom72
Predictor        Coef      SE Coef         T        P
Constant       -19.02        13.20     -1.44    0.153
Comper        -26.393        9.890     -2.67    0.009
Homper        -12.123        7.508     -1.61    0.110
Indper        -15.531        8.630     -1.80    0.075
sizehse         7.219        2.138      3.38    0.001
incom72      0.004081     0.001555      2.62    0.010

S = 3.949       R-Sq = 40.1%      R-Sq(adj) = 36.5%
Analysis of Variance
Source           DF          SS        MS         F        P
Regression        5      876.80    175.36     11.25    0.000
Residual Error   84     1309.83     15.59
Total            89     2186.63

Source       DF      Seq SS
Comper        1      245.47
Homper        1        1.38
Indper        1      112.83
sizehse       1      409.77
incom72       1      107.36

Unusual Observations
Obs     Comper     hseval        Fit      SE Fit    Residual    St Resid
 23      0.100     20.003     28.296       1.913      -8.294     -2.40RX
 24      0.103     20.932     29.292       2.487      -8.360     -2.73RX
 29      0.139     16.498     19.321       1.872      -2.823     -0.81 X
 30      0.141     16.705     19.276       1.859      -2.570     -0.74 X
 75      0.112     35.976     24.513       0.747      11.463      2.96R
 76      0.116     35.736     24.418       0.749      11.317      2.92R
R denotes an observation with a large standardized residual
X denotes an observation whose X value gives it large influence.
Durbin-Watson statistic = 1.03
```

Dropping the insignificant independent variables: Homper and Indper yields:

Regression Analysis: hseval versus Comper, sizehse, incom72

```
The regression equation is
hseval = - 34.2 - 13.9 Comper + 8.27 sizehse + 0.00364 incom72

Predictor        Coef      SE Coef        T       P
Constant       -34.24        10.44    -3.28   0.002
Comper        -13.881        6.974    -1.99   0.050
sizehse         8.270        1.957     4.23   0.000
incom72      0.003636     0.001456     2.50   0.014

S = 3.983       R-Sq = 37.6%      R-Sq(adj) = 35.4%
Analysis of Variance
Source            DF        SS        MS        F       P
Regression         3    822.53    274.18    17.29   0.000
Residual Error    86   1364.10     15.86
Total             89   2186.63

Source        DF      Seq SS
Comper         1      245.47
sizehse        1      478.09
incom72        1       98.98

Unusual Observations
Obs     Comper     hseval       Fit    SE Fit   Residual   St Resid
 49      0.282     29.810    23.403     1.576      6.407     1.75 X
 50      0.284     30.061    23.380     1.583      6.681     1.83 X
 75      0.112     35.976    24.708     0.674     11.268     2.87R
 76      0.116     35.736    24.659     0.667     11.077     2.82R

R denotes an observation with a large standardized residual
X denotes an observation whose X value gives it large influence.
Durbin-Watson statistic = 1.02
```

Excluding median rooms per residence (Sizehse):

Regression Analysis: hseval versus Comper, incom72

```
The regression equation is
hseval = 4.69 - 20.4 Comper + 0.00585 incom72
Predictor        Coef      SE Coef        T       P
Constant        4.693        5.379     0.87   0.385
Comper        -20.432        7.430    -2.75   0.007
incom72      0.005847     0.001484     3.94   0.000

S = 4.352       R-Sq = 24.7%      R-Sq(adj) = 22.9%
Analysis of Variance
Source            DF        SS        MS        F       P
Regression         2    539.20    269.60    14.24   0.000
Residual Error    87   1647.44     18.94
Total             89   2186.63
Source        DF      Seq SS
Comper         1      245.47
incom72        1      293.73
Durbin-Watson statistic = 0.98
```

Note that the coefficient on percent of commercial property for both of the models is negative; however, it is larger in the second model where the median rooms variable is excluded.

13.24 If y is, in fact, strongly influenced by x_2, dropping it from the regression equation could lead to serious specification bias. Instead of dropping the variable, it is preferable to acknowledge that, while the group as a whole is clearly influential, the data does not contain information to allow the disentangling of the separate effects of each of the explanatory variables with some degree of precision.

13.26 a. Graphical check for heteroscedasticity shows no evidence of strong heteroscedasticity.

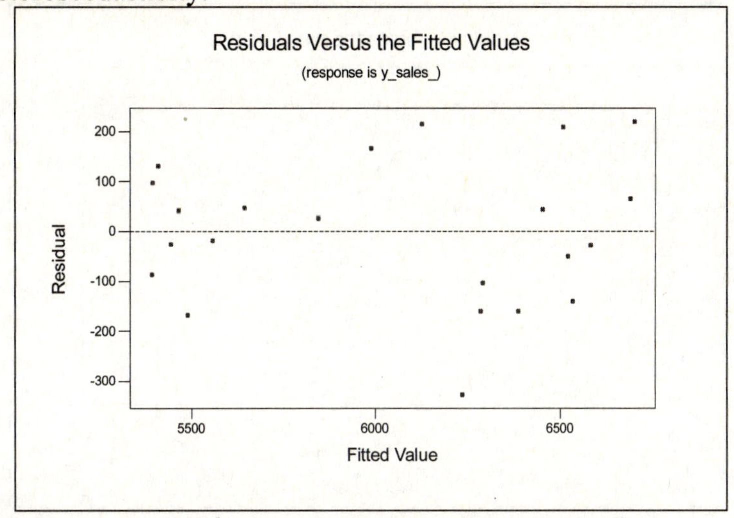

b. The auxiliary regression is $\hat{e}^2 = -63310.41 + 13.75\hat{y}$

$n = 22,\quad R^2 = .06954,\quad nR^2 = 1.5299 < 2.71 = \chi^2_{1,.1}$

therefore, do not reject H_0 the error terms have constant variance at the 10% level.

13.28 a. Compute the multiple regression of Y on x_1, x_2 and x_3.

Results for: Household Income.MTW

Regression Analysis: y versus X1, X2, X3

```
The regression equation is
y = 0.2 + 0.000406 X1 + 4.84 X2 - 1.55 X3
Predictor        Coef     SE Coef      T      P  VIF
Constant         0.16       34.91   0.00  0.996
X1          0.0004060   0.0001736   2.34  0.024  1.2
X2              4.842       2.813   1.72  0.092  1.5
X3            -1.5543      0.3399  -4.57  0.000  1.3

S = 3.04752   R-Sq = 54.3%   R-Sq(adj) = 51.4%
Analysis of Variance
Source          DF      SS      MS      F      P
Regression       3  508.35  169.45  18.24  0.000
Residual Error  46  427.22    9.29
Total           49  935.57

Source  DF  Seq SS
X1       1  157.43
X2       1  156.76
X3       1  194.15

Unusual Observations
Obs     X1       y     Fit  SE Fit  Residual  St Resid
  4  22456  52.600  59.851   0.850    -7.251    -2.48R
 13  14174  44.200  50.840   1.166    -6.640    -2.36R
R denotes an observation with a large standardized residual.
Durbin-Watson statistic = 1.75105
```

b. Graphical check for heteroscedasticity shows no evidence of strong heteroscedasticity

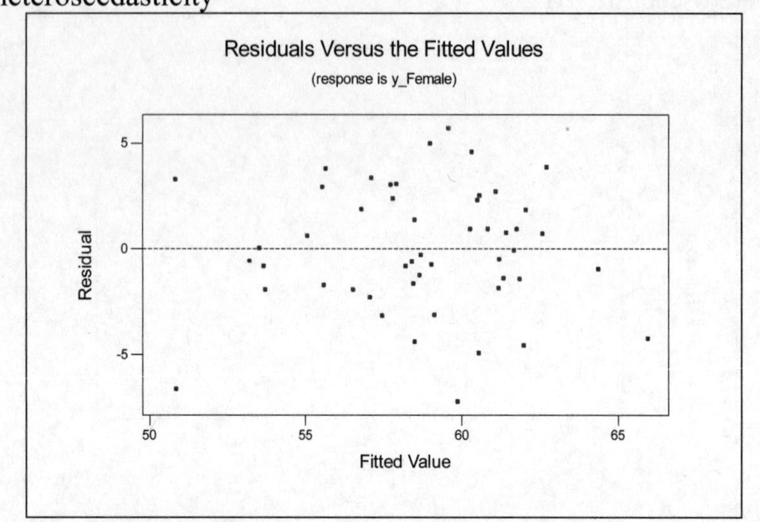

c. The auxiliary regression is $e^2 = 20.34 - .201\hat{y}$

$n = 50,\quad R^2 = .00322,\ nR^2 = .161 < 2.71 = \chi^2_{1,.1}$, therefore, do not reject the H_0 that the error terms have constant variance at the 10% level.

13.30 Test for the presence of autocorrelation.
$H_0 : \rho = 0, H_1 : \rho > 0$, d = .50, n = 30, K = 3, $\alpha = .05$: $d_L = 1.21$ and $d_U = 1.65$
$\alpha = .01$: $d_L = 1.01$ and $d_U = 1.42$
Reject the null hypothesis based on the Durbin-Watson test at both the 5% and 1% levels. Estimate of the autocorrelation coefficient: $r = 1 - \frac{d}{2} =$
$r = 1 - \frac{.5}{2} = .75$

a. $H_0 : \rho = 0, H_1 : \rho > 0$, d = .80, n = 30, K = 3, $\alpha = .05$: $d_L = 1.21$ and $d_U = 1.65$
$\alpha = .01$: $d_L = 1.01$ and $d_U = 1.42$
Reject the null hypothesis based on the Durbin-Watson test at both the 5% and 1% levels. Estimate of the autocorrelation coefficient:
$r = 1 - \frac{d}{2} = r = 1 - \frac{.8}{2} = .60$

b. $H_0 : \rho = 0, H_1 : \rho > 0$, d = 1.10, n = 30, K = 3, $\alpha = .05$: $d_L = 1.21$ and $d_U = 1.65$
$\alpha = .01$: $d_L = 1.01$ and $d_U = 1.42$
Reject the null hypothesis based on the Durbin-Watson test at the 5% level. The test is inconclusive at the 1% level.
Estimate of the autocorrelation coefficient: $r = 1 - \frac{d}{2} = r = 1 - \frac{1.10}{2} = .45$

c. $H_0: \rho = 0, H_1: \rho > 0$, d = 1.25, n = 30, K = 3, $\alpha = .05$: $d_L = 1.21$ and $d_U = 1.65$
$\alpha = .01$: $d_L = 1.01$ and $d_U = 1.42$
The test is inconclusive at both the 5% level and the 1% level.

d. $H_0: \rho = 0, H_1: \rho > 0$, d = 1.70, n = 30, K = 3, $\alpha = .05$: $d_L = 1.21$ and $d_U = 1.65$
$\alpha = .01$: $d_L = 1.01$ and $d_U = 1.42$
Do not reject the null hypothesis at either the 5% level or the 1% level. There is insufficient evidence to suggest autocorrelation exists in the residuals.

13.32 a. $n = 30, \quad R^2 = .043$

$nR^2 = 1.29 < 2.71 = \chi^2_{1,.1}$, therefore, do not reject the H_0 that the error terms have constant variance at the 10% level.

b. $H_0: \rho = 0, H_1: \rho > 0$, d=1.29, n=30, K = 4, $\alpha = .05$: $d_L = 1.14$ and $d_U = 1.74$
$\alpha = .01$: $d_L = .94$ and $d_U = 1.51$
The Durbin-Watson test gives inconclusive results at both the 5% and 1% levels.

13.34 $n = 27, \quad R^2 = .087$, $nR^2 = 2.349 < 2.71 = \chi^2_{1,.1}$, therefore, do not reject H_0 that the error terms have constant variance at the 10% level.

13.36 The regression model associated with Exercise 13.18 includes the lagged value of the dependent variable as an independent variable. In the presence of a lagged dependent variable used as an independent variable, the Durbin-Watson statistic is no longer valid. Instead, use of Durbin's h statistic is appropriate:

$H_0: \rho = 0, H_1: \rho > 0$

$$r = 1 - \frac{d}{2} = 1 - \frac{1.63}{2} = .185, \quad s_c^{\,2} = (.1243)^2 = .01545$$

$$h = r\sqrt{\frac{n}{1 - n(s_c^{\,2})}} = .185\sqrt{\frac{28}{1 - 28(.01545)}} = 1.30$$

$z_{.1} = 1.28$, therefore, reject H_0 at the 10% level but not at the 5% level

13.38 a. **Results for: Advertising Retail.xls**
Regression Analysis: Retail Sales X(t) versus Advertising Y(t)

```
The regression equation is
Retail Sales X(t) = 2269 + 28.5 Advertising Y(t)

Predictor        Coef     SE Coef         T       P
Constant       2269.5       278.4      8.15   0.000
Advertis       28.504       2.087     13.66   0.000

S = 161.7        R-Sq = 90.3%     R-Sq(adj) = 89.8%
Analysis of Variance
Source            DF        SS         MS         F        P
Regression         1   4874833    4874833    186.52    0.000
Residual Error    20    522728      26136
Total             21   5397561
Durbin-Watson statistic = 1.12
```

b. $H_0: \rho = 0, H_1: \rho > 0$
d = 1.12, n=22, K = 1
$\alpha = .05$: $d_L = 1.24$ and $d_U = 1.43$
$\alpha = .01$: $d_L = 1.00$ and $d_U = 1.17$
Reject H_0 at the 5% level, autocorrelation of the residuals exists at the 5% level, test is inconclusive at the 1% level

c. The fitted regression is
$$y_t - .5597 y_{t-1} = 25.6122(1 - .5597) + 1.4482(x_t - .5597 x_{t-1})$$

13.40 $H_0: \rho = 0, H_1: \rho > 0$
Note that due to the presence of a lagged dependent variable used as an independent variable, Durbin's h statistic is relevant
$$r = 1 - \frac{d}{2} = 1 - \frac{1.82}{2} = .09, \quad s_c^2 = (.136)^2 = .0185$$
$$h = r\sqrt{\frac{n}{1 - n(s_c^2)}} = .09\sqrt{\frac{25}{1 - 25(.0185)}} = .61374$$
$z_{.1} = 1.28$, therefore, do not reject H_0 at the 10% level

13.42 In the first case, the coefficients of the dummy variables measures the difference between the expected tax revenues (as a percentage of gross national product) in the countries that participate in some form of economic integration versus those that do not. It quantifies the difference between the value of 1 of the dummy variable versus the excluded category.
In the second case, the intercept terms will include the effect of participation by a country in some form of economic integration on tax revenues.

13.44 a. Heteroscedasticity: is defined as when the residuals do not have constant variance at all levels of the dependent variable. It results in parameter estimates that are not efficient and invalidates the confidence interval and hypothesis testing statistics

b. Autocorrelated errors: Autocorrelation of the residuals is when the error terms are not independent from one another across the order of observation. It results in inefficient parameter estimates and invalidates the confidence interval and hypothesis testing statistics. For models that contain lagged dependent variables, autocorrelated errors will result in inconsistent parameter estimates

13.46 a. All else equal, the secondary market price of $100 of debt of a country is 9.6 units lower if U.S. bank regulators have mandated write-downs of the country's assets than if otherwise.

b. $H_0 : \beta_3 = 0, \quad H_1 : \beta_3 < 0$

$$t = \frac{-.15}{.056} = -2.679$$

Reject H_0 at the 1% level since $t < -2.5 = t_{23,.01}$

c. 84% of the variability in the secondary market price of $100 of debt of a country is explained by the variability in each of the independent variables

d. Any improvement or change in the model characteristics would have to be weighed against the loss of the degrees of freedom from an added independent variable

13.48 a. All else equal, a one unit increase in the world price of U.S. wheat will yield an estimated decrease of .62 of a unit in the quantity of U.S. wheat exported.

b. $H_0 : \beta_3 = 0, \quad H_1 : \beta_3 > 0, \; t = \frac{.61}{.21} = 2.905$

Reject H_0 at the .5% level since $t > 2.771 = t_{27,.005}$

c. $H_0 : \rho = 0, H_1 : \rho > 0$, d = .61, n=32, K = 4, $\alpha = .01$: $d_L = .98$ and $d_U = 1.51$
Reject H_0 at the 1% level

d. Given that the residuals are autocorrelated, the hypothesis test results of part b are not valid. The model must be reestimated taking into account the autocorrelated errors

13.50 a. All else equal, a bank whose head office is in New York will experience a 1.67% higher rate of return than one which is based outside of New York

b. $H_0 : \beta_1 = \beta_2 = \beta_3 = \beta_4 = 0, H_1 : \textit{at least one } \beta_i \neq 0 \; (i = 1,2,3,4)$

$F = \frac{25}{4} \frac{.317}{1 - .317} = 2.901, \quad F_{4,25,.05} = 2.76$, therefore, reject H_0 at the 5% level

c. $nR^2 = 30(.082) = 2.46 < 2.71 = \chi^2_{1,.1}$, therefore, do not reject H_0 that the error terms have constant variance at the 10% level.

13.52 a. All else equal, an additional room in a dwelling leads to an expected \$10.94 increase in the average monthly electrical bill

b. $H_0 : \beta_1 = 0, \quad H_1 : \beta_1 \neq 0$, $t = -.956$

Do not reject H0 at the 20% level since $t = -.956 > -1.321 = t_{22,.1}$

c. An indication of the presence of multicollinearity occurs when, taken as a group, the set of independent variables appears to exert a considerable influence on the dependent variable, but when looked at separately through hypothesis tests, none of the individual regression slope coefficients appear significantly different from zero. So if the F-test of the significance of the overall regression show the model has significant explanatory power and yet none of the individual regression slope coefficients is significant, suspect that multicollinearity exists between the independent variables.

d. If an important independent variable is omitted from the regression model, the least squares estimates will be unreliable due to specification bias.

e. $nR^2 = 25(.047) = 1.175 < 2.71 = \chi^2_{1,.1}$, therefore, do not reject the null hypothesis that the error terms have constant variance at the 10% level

13.54 **Regression Analysis: y versus x1, x2**

```
The regression equation is
y = - 1.63 + 0.779 x1 + 1.50 x2
Predictor        Coef       SE Coef          T        P
Constant       -1.627         1.130      -1.44    0.166
x1             0.7792        0.1649       4.72    0.000
x2             1.4961        0.1802       8.30    0.000

S = 1.435        R-Sq = 80.2%      R-Sq(adj) = 78.1%
Analysis of Variance
Source            DF          SS          MS          F        P
Regression         2     158.611      79.305      38.50    0.000
Residual Error    19      39.142       2.060
Total             21     197.753

Source       DF        Seq SS
x1            1        16.626
x2            1       141.985
Durbin-Watson statistic = 1.51
```

Test for heteroscedasticity:

$nR^2 = 22(.02215) = .4873 < 2.71 = \chi^2_{1,.1}$, therefore, do not reject H0 that the error terms have constant variance at the 10% level

Test for autocorrelation:

$H_0 : \rho = 0, H_1 : \rho > 0$

d = 1.51, n=22, K = 2

$\alpha = .05$: $d_L = 1.15$ and $d_U = 1.54$

$\alpha = .01$: $d_L = .91$ and $d_U = 1.28$

At the 5% level the Durbin-Watson test is inconclusive. Do not reject H0 at the 1% level

13.56 76.6% of the variation in the FDIC examiner work hours can be explained by the variation in total assets of the bank, total number of offices, classified to total loan ratio, management rating and if the examination was conducted jointly with the state.

$H_0: \beta_1 = \beta_2 = \cdots = \beta_7 = 0,\quad H_1: at\ least\ one\ \beta_i \neq 0\ (i = 1,...,7)$

$F = \frac{83}{7}\frac{.766}{1-.766} = 38.814$, $F_{7,83,.01} \approx 2.95$, therefore, reject H_0 at the 1% level

13.58 a. **Regression Analysis: Nonresidential versus Prime Rate**

```
The regression equation is
Nonresidential = 1282 - 53.0 Prime Rate

Predictor        Coef  SE Coef      T      P
Constant      1282.08    69.68  18.40  0.000
Prime Rate    -53.011    7.278  -7.28  0.000

S = 258.392   R-Sq = 32.3%   R-Sq(adj) = 31.7%

Analysis of Variance

Source           DF        SS       MS      F      P
Regression        1   3542613  3542613  53.06  0.000
Residual Error  111   7411062    66766
Total           112  10953675

Durbin-Watson statistic = 0.05
```

Test for autocorrelation:

$H_0: \rho = 0, H_1: \rho > 0$, d = 0.05, n=113, K = 1, $\alpha = .05$: $d_L = 1.65$ and $d_U = 1.69$

$\alpha = .01$: $d_L = 1.52$ and $d_U = 1.56$, reject H_0 at the 1% level or 5% level.

b. **Regression Analysis**

```
The regression equation is
Nonresidential = - 76.7 + 3.94 Prime Rate + 0.207 Gross
domestic product
           - 0.613 Federal - 0.902 State and local
           + 0.0299 Per capita Income (Lagged)

Predictor         Coef        StDev          T          P
Constant        -76.70        15.63      -4.91      0.000
Prime Ra         3.938        1.421       2.77      0.007
Gross do       0.20697      0.01963      10.54      0.000
Federal       -0.61311      0.09838      -6.23      0.000
State an       -0.9023       0.1786      -5.05      0.000
Per capi      0.029940     0.007895       3.79      0.000

S = 13.97         R-Sq = 95.2%      R-Sq(adj) = 95.0%

Analysis of Variance

Source            DF          SS          MS         F         P
Regression         5      412868       82574    422.96     0.000
Residual Error   106       20694         195
Total            111      433562

Durbin-Watson statistic = 1.59
```

Test for autocorrelation:
$H_0: \rho = 0, H_1: \rho > 0$, d = 1.59, n=112, K = 5, $\alpha = .05$: d_L = 1.57 and d_U = 1.78
$\alpha = .01$: d_L = 1.44 and d_U = 1.65, test is inconclusive at the 1% level and 5% level.

c. The regression with prime rate had a $r^2 = 0.05$ and high autocorrelation. It explained only 5% of the variation. The corrected for autocorrelation regression equation using all the predictor variables had a $r^2 = 0.952$ and no autocorrelation. It explained 95.2% of the variation. The second regression model explains the investment much better.

13.60 a. Regression of value added as a function of inputs labor and capital is as follows:

Regression Analysis: valadded versus labor, capital

```
The regression equation is
valadded = 123 + 2.32 labor + 0.472 capital
Predictor        Coef     SE Coef          T        P
Constant        122.7       170.9       0.72    0.480
labor           2.323       1.033       2.25    0.034
capital        0.4716      0.1123       4.20    0.000

S = 469.9       R-Sq = 96.0%      R-Sq(adj) = 95.6%
Analysis of Variance
Source            DF         SS          MS          F        P
Regression         2  126519178    63259589     286.46    0.000
Residual Error    24    5299991      220833
Total             26  131819169

Source       DF      Seq SS
labor         1   122623731
capital       1     3895447
Durbin-Watson statistic = 2.02
```

b. Plot of residuals versus labor and capital is as follows:

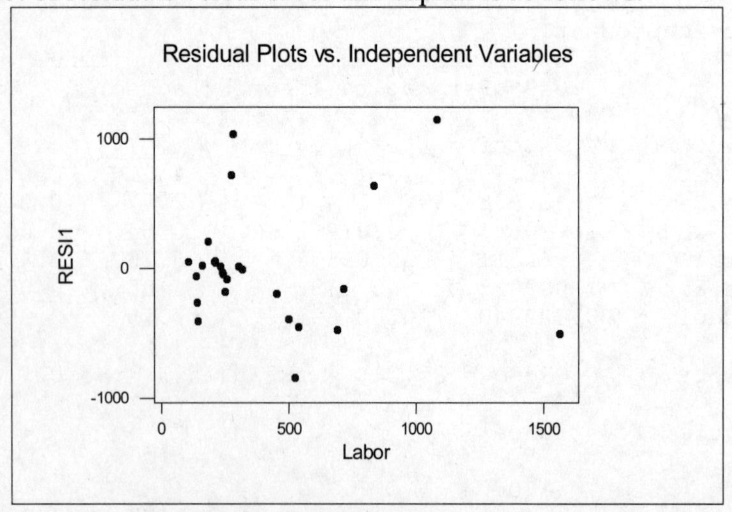

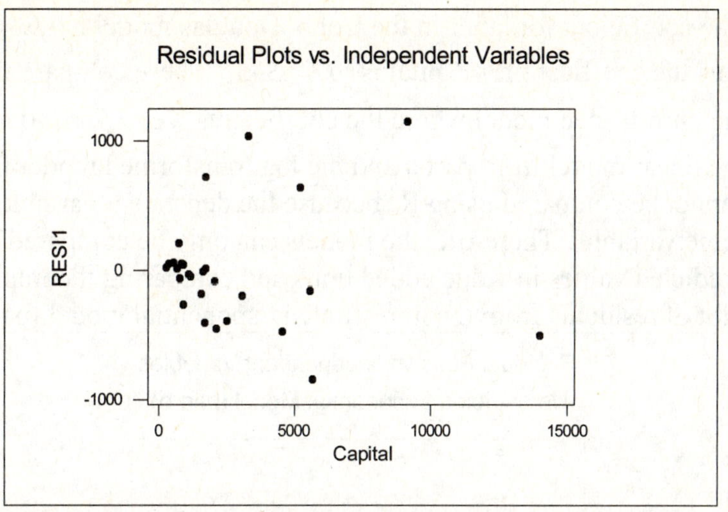

Residual plots indicate possible evidence of increasing variance.

c. Estimate the Cobb–Douglas production function.

Regression Analysis: lnvaladd versus lnlabor, lncapital

```
The regression equation is
lnvaladd = 1.29 + 0.596 lnlabor + 0.366 lncapital

Predictor        Coef    SE Coef         T       P
Constant       1.2883     0.3225      3.99   0.001
lnlabor        0.5960     0.1293      4.61   0.000
lncapita      0.36638    0.09009      4.07   0.000

S = 0.1918      R-Sq = 94.1%      R-Sq(adj) = 93.7%

Analysis of Variance

Source            DF        SS        MS       F       P
Regression         2   14.1803    7.0902  192.74   0.000
Residual Error    24    0.8829    0.0368
Total             26   15.0632

Durbin-Watson statistic = 1.68
```

The coefficient for labor is approximately .60 and the coefficient for capital is roughly .37. Sum is .97, which is close to the restriction implied by the Cobb–Douglas model.

d. Estimate the Cobb–Douglas production function with constant returns to scale.

Regression Analysis: lnvaldif versus lnlabdif

```
The regression equation is
lnvaldif = 1.11 + 0.657 lnlabdif
Predictor        Coef    SE Coef         T       P
Constant       1.1120     0.1410      7.89   0.000
lnlabdif      0.65723    0.08031      8.18   0.000

S = 0.1894      R-Sq = 72.8%      R-Sq(adj) = 71.7%

Analysis of Variance

Source            DF        SS        MS       F       P
Regression         1    2.4015    2.4015   66.97   0.000
Residual Error    25    0.8965    0.0359
Total             26    3.2981

Durbin-Watson statistic = 1.67
```

The coefficient for labor in the Cobb–Douglas model is, $\beta_1 = .657$ and thus the coefficent for capital is $\beta_2 = .343$. These compare favorably to the unrestricted model where the coefficients were $\beta_1 = .60$ *and* $\beta_2 = .37$.

e. The linear model from part a and the log transformed model from part b cannot be compared using R^2 because the dependent variables are not the same variable. Therefore, the models can only be compared by computing predicted values in value added units and comparing the implied residuals. Plot of residuals from the unrestricted exponential model (part b):

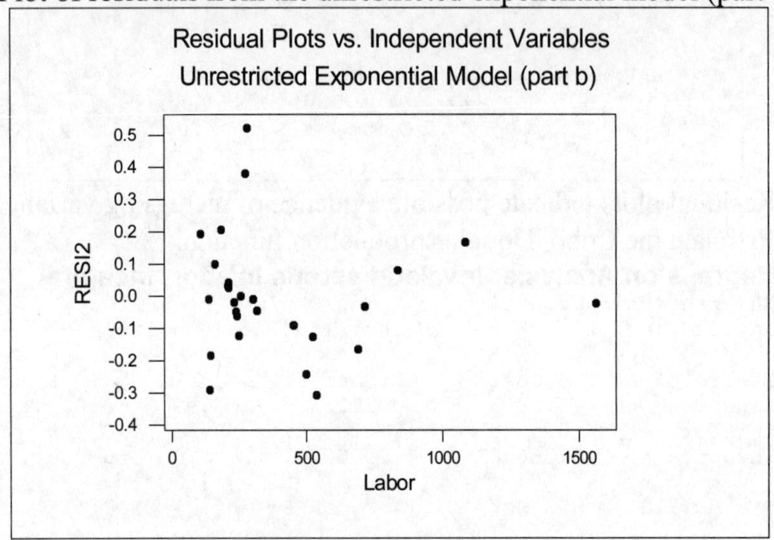

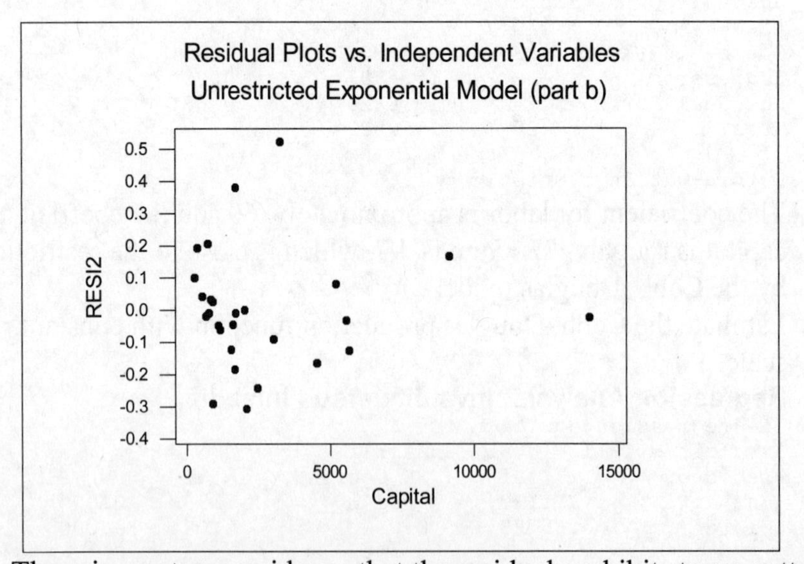

There is no strong evidence that the residuals exhibit strong patterns. However, there are two quite large positive residuals with values around 1,400, which is almost twice that of the next largest positive residuals.

Plot the residuals from the restricted exponential model (part c):

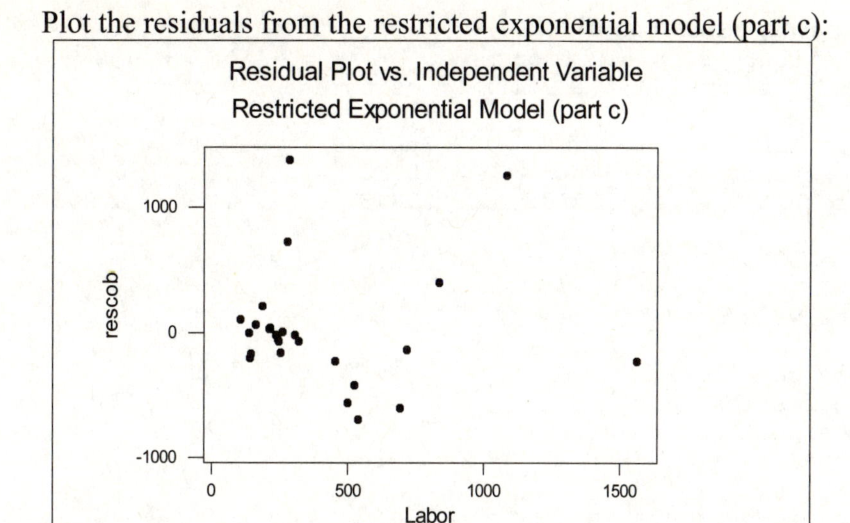

No evidence of strong data patterns, may want to check for increasing variance.

Descriptive Statistics: linresid, reslnmd, rescob

Variable	N	Mean	Median	TrMean	StDev	SE Mean
linresid	27	-0.0	-24.0	-12.6	451.5	86.9
reslnmd	27	58.6	-22.4	28.0	486.7	93.7
rescob	27	22.0	-18.7	-3.7	472.4	90.9

Variable	Minimum	Maximum	Q1	Q3
linresid	-841.4	1155.9	-263.6	56.0
reslnmd	-619.5	1503.8	-202.5	90.0
rescob	-691.1	1379.2	-201.7	62.4

Compare the standard deviations of the implied residuals:

Descriptive Statistics: linresid, reslnmd, rescob

Variable	N	Mean	Median	TrMean	StDev	SE Mean
linresid	27	-0.0	-24.0	-12.6	451.5	86.9
reslnmd	27	58.6	-22.4	28.0	486.7	93.7
rescob	27	22.0	-18.7	-3.7	472.4	90.9

Variable	Minimum	Maximum	Q1	Q3
linresid	-841.4	1155.9	-263.6	56.0
reslnmd	-619.5	1503.8	-202.5	90.0
rescob	-691.1	1379.2	-201.7	62.4

Note that both the mean and standard deviation of the implied residuals from the restricted exponential model (part c) are smaller than from the unrestricted exponential model (part b).

13.62 a. **Regression Analysis**

The regression equation is
Services = - 74.2 + 0.411 Gross domestic product

Predictor	Coef	StDev	T	P
Constant	-74.16	15.28	-4.85	0.000
Gross do	0.410936	0.001843	223.03	0.000

S = 39.09   R-Sq = 99.8%   R-Sq(adj) = 99.8%

Analysis of Variance

Source	DF	SS	MS	F	P
Regression	1	75989723	75989723	49742.82	0.000
Residual Error	111	169569	1528		
Total	112	76159292			

Durbin-Watson statistic = 0.26

Test for autocorrelation:

$H_0: \rho = 0, H_1: \rho > 0$, d = 0.26, n=113, K = 1, $\alpha = .05$: d_L = 1.65 and d_U = 1.69

$\alpha = .01$: d_L = 1.52 and d_U = 1.56, reject H_0 at the 1% level and 5% level. There is autocorrelation.

b. **Regression Analysis**

The regression equation is
Services = 39.4 + 0.288 Gross domestic product
+ 0.218 Personal consumption (lagged) - 1.48 Imports of Services- 8.26 Prime Rate

Predictor	Coef	StDev	T	P
Constant	39.39	45.90	0.86	0.393
Gross do	0.28837	0.02265	12.73	0.000
Personal	0.21841	0.03665	5.96	0.000
Imports	-1.4819	0.4097	-3.62	0.000
Prime Ra	-8.259	1.164	-7.09	0.000

S = 29.13   R-Sq = 99.9%   R-Sq(adj) = 99.9%

Analysis of Variance

Source	DF	SS	MS	F	P
Regression	4	76067648	19016912	22410.88	0.000
Residual Error	108	91644	849		
Total	112	76159292			

Durbin-Watson statistic = 0.40

Test for autocorrelation:

$H_0: \rho = 0, H_1: \rho > 0$, d = 0.40, n=113, K = 4, $\alpha = .05$: d_L = 1.59 and d_U = 1.76

$\alpha = .01$: d_L = 1.46 and d_U = 1.63, reject H_0 at the 1% level and 5% level. There is still autocorrelation. The new model reduces but does not eliminate autocorrelation.

Chapter 14:

Analysis of Categorical Data

14.2 H_0: Mutual fund performance is equally likely to be in the 5 performance quintiles.
H_1: otherwise

Mutual funds	Top 20%	2nd 20%	3rd 20%	4th 20%	5th 20%	Total
Observed Number	13	20	18	11	13	75
Probability (Ho)	0.2	0.2	0.2	0.2	0.2	1
Expected Number	15	15	15	15	15	75
Chi-square calculation	0.266667	1.666667	0.6	1.066667	0.266667	3.8667

Chi-square calculation: $\chi^2 = \sum \frac{(O_i - E_i)^2}{Ei}$ = 3.8667

$\chi^2_{(4,.1)} = 7.78$ Therefore, fail to reject H_0 at the 10% level

14.4 H_0: Quality of the output conforms to the usual pattern
H_1: otherwise

Electronic component	No faults	1 fault	>1 fault	Total
Observed Number	458	30	12	500
Probability (Ho)	0.93	0.05	0.02	1
Expected Number	465	25	10	500
Chi-square calculation	0.105376344	1	0.4	1.505376

Chi-square calculation: $\chi^2 = \sum \frac{(O_i - E_i)^2}{Ei}$ = 1.505

$\chi^2_{(2,.05)} = 5.99$ Therefore, do not reject H_0 at the 5% level

14.6 H_0: Student opinion of business courses is the same as that for all courses
H_1: otherwise

Opinion	Very useful	Somewhat	Worthless	Total
Observed Number	68	18	14	100
Probability (Ho)	0.6	0.2	0.2	1
Expected Number	60	20	20	100
Chi-square calculation	1.066666667	0.2	1.8	3.066667

Chi-square calculation: $\chi^2 = \sum \frac{(O_i - E_i)^2}{Ei}$ = 3.067

$\chi^2_{(2,.10)} = 4.61$ Therefore, do not reject H_0 at the 10% level

14.8 H_0: Consumer preferences for soft drinks are equally spread across 5 soft drinks

H_1: otherwise

Drink	A	B	C	D	E	Total
Observed Number	20	25	28	15	27	115
Probability (Ho)	0.2	0.2	0.2	0.2	0.2	1
Expected Number	23	23	23	23	23	115
Chi-square calculation	0.391304	0.173913	1.086957	2.782609	0.695652	5.130435

Chi-square calculation: $\chi^2 = \sum \frac{(O_i - E_i)^2}{Ei}$ = 5.130

$\chi^2_{(4,.10)} = 7.78$ Therefore, do not reject H_0 at the 10% level

14.10 H_0: Statistics professors preferences for software packages are equally divided across 4 packages

H_1: otherwise

Software	M	E	S	P	Total
Observed Number	100	80	35	35	250
Probability (Ho)	0.25	0.25	0.25	0.25	1
Expected Number	62.5	62.5	62.5	62.5	250
Chi-square calculation	22.5	4.9	12.1	12.1	51.6

Chi-square calculation: $\chi^2 = \sum \frac{(O_i - E_i)^2}{Ei}$ = 51.6

$\chi^2_{(3,.005)} = 12.84$ Therefore, reject H_0 at the .5% level

14.12 H_0: population distribution of arrivals per minute is Poisson

H_1: otherwise

Arrivals	0	1	2	3	4+	Total
Observed Number	10	26	35	24	5	100
Probability (Ho)	0.1496	0.2842	0.27	0.171	0.1252	1
Expected Number	14.96	28.42	27	17.1	12.52	100
Chi-square calculation	1.644492	0.206066	2.37037	2.784211	4.516805	11.52194

Chi-square calculation: $\chi^2 = \sum \frac{(O_i - E_i)^2}{Ei}$ = 11.52

$\chi^2_{(3,.01)}$=11.34 $\chi^2_{(3,.005)}$=12.84. Reject H_0 at the 1% level but not at the .5% level

14.14 H_0: resistance of electronic components is normally distributed

H_1: otherwise

$$B = 100\left[\frac{(.63)^2}{6} + \frac{(3.85-3)^2}{24}\right] = 9.625$$

From Table 14.9 – Significance points of the Jarque-Bera statistic; 5% point (n=100) is 4.29. Therefore, reject H_0 at the 5% level

14.16 H_0: monthly balances for credit card holders of a particular card are normally distributed

H_1: otherwise

$$B = 125\left[\frac{(.55)^2}{6} + \frac{(2.77-3)^2}{24}\right] = 6.578$$

From Table 14.9 – Significance points of the Jarque-Bera statistic; 5% point (n = 125) is 4.34. Therefore, reject H_0 at the 5% level

14.18 H_0: No association exists between gpa and major

H_1: otherwise

Chi-Square Test: GPA<3, GPA3+

Expected counts are printed below observed counts

	GPA<3	GPA3+	Total
1	50	35	85
	46.75	38.25	
2	45	30	75
	41.25	33.75	
3	15	25	40
	22.00	18.00	
Total	110	90	200

Chi-Sq = 0.226 + 0.276 + 0.341 + 0.417 + 2.227 + 2.722 = 6.209

DF = 2, P-Value = 0.045

$\chi^2_{(2,.05)} = 5.99$ Therefore, reject H_0 of no association at the 5% level

14.20 a. Complete the contingency table:

		Method of learning about product	
Age	Friend	Ad	col. total
<21	30	20	50
21-35	60	30	90
35+	18	42	60
row total	108	92	200

b. H_0: No association exists between the method of learning about the product and the age of the respondent

H_1: otherwise

Chi-Square Test: Friend, Ad

Expected counts are printed below observed counts

	Friend	Ad	Total
1	30	20	50
	27.00	23.00	
2	60	30	90
	48.60	41.40	
3	18	42	60
	32.40	27.60	
Total	108	92	200

Chi-Sq = 0.333 + 0.391 +2.674 + 3.139 +6.400 + 7.513 = 20.451

DF = 2, P-Value = 0.000

$\chi^2_{(2,.005)} = 10.6$ Therefore, reject H_0 of no association at the .5% level

14.22 $H_0: P = 0.50$ (there is no preference for one stock over the other)

$H_1: P \neq 0.50$ (otherwise)

n = 11 For stock 2 and a two-sided test, $P(2 \geq X \geq 9) = 2P(X \leq 1) =$
2[.0005 + .0054] = .0118

Therefore, reject H_0 at levels of alpha in excess of 1.18%

14.24 $H_0: P = 0.50$ (grocery store managers are equally divided about customers attitudes about electronic coupons)

$H_1: P \neq 0.50$ (otherwise)

n = 11. For 8 "yes" answers and a two-sided test,
$P(4 \leq X \leq 7) = 2P(X \leq 3) = 2[.0005 + .0054 + .0269 + .0806] = .2268$

Therefore, reject H_0 at levels of alpha in excess of 22.68%

14.26 $H_0: P = 0.50$ (voters are evenly divided)

$H_1: P \neq 0.50$ (otherwise)

n = 130 – 18 = 112. $\hat{p} = 68/112 = .6071$

$\mu = nP = 112(.5) = 56 \quad \sigma = .5\sqrt{n} = .5\sqrt{112} = 5.2915$

$Z = \frac{S^* - \mu}{\sigma} = \frac{67.5 - 56}{5.2915} = 2.17$, p-value = $2[1 - F_z(2.17)] = 2[1 - .9850] = .030$

Therefore, reject H_0 at levels of alpha in excess of .30%

14.28 Open-ended question. The findings should include statements about the relative size of the firms. The MIPS firms have larger total assets than do non-MIPS comparable firms. This holds true in both the Utilities as well as Industrial industries and for the overall total. Results of interest coverage and long-term debt-to-total-asset ratios varies depending on which test and which type of industry the firms are in. While publicly traded MIPS firms in the utilities industries have significantly higher long-term debt-to-total-asset ratio then do non-MIPS firms, the MIPS firms in the industrials do not.

14.30 H_0: origin of the student has no effect on GPA

H_1: otherwise

Wilcoxon Signed Rank Test: Diff_14.30

Test of median = 0.000000 versus median not = 0.000000

	N	N for Test	Wilcoxon Statistic	P	Estimated Median
Diff_14.30	8	7	18.5	0.499	0.1000

$n = 7, \quad T = 9.5, \quad T_{.10} = 6$. Do not reject H_0 at any common level of alpha

14.32 H_0: Total quality management has no impact on job satisfaction of employees

H_1: Total quality management has a positive impact on job satisfaction

$T = 169, \quad \mu_T = 232.5, \quad \sigma^2{}_T = 30(31)(61)/24 = 2363.75$

$z = \dfrac{169 - 232.5}{\sqrt{2363.75}} = -1.31$, p-value = 1-F$_Z$(1.31) = 1- .9049 = .0951

Therefore, reject H_0 at levels in excess of 9.51%

14.34 H_0: directors' ownership is the same for firms with and without an audit committee

H_1: directors' ownership is higher for firms without an audit committee

$z = \dfrac{U - \mu_U}{\sigma_U} = 2.01$, p-value = 1-F$_Z$(2.01) = 1- .9778 = .0222

Therefore, reject H_0 at levels in excess of 2.22%

14.36 H_0: starting salaries are equal for both types of schools

H_1: otherwise (two-tailed)

Sum of ranks for technical college = 153.5

$R_1 = 153.5, n_1 = 12, n_2 = 10 \qquad U = 12(10) + 12(13)/2 - 153.5 = 44.5$

$\mu_U = 120/2 = 60, \sigma^2{}_U = 120(23)/12 = 230$

$z = \dfrac{44.5 - 60}{\sqrt{230}} = -1.02$, p-value = 2[1-F$_Z$(1.02)] = 2[1-.8461] = .3078

Therefore, reject H_0 at levels in excess of 30.78%

14.38 H_0: no difference in rankings by gender

H_1: otherwise

Sum of ranks for males = 242

$T = 242, n_1 = 15, n_2 = 15, \ E(T) = \mu_T = n_1(n_1 + n_2 + 1)/2 = 232.5$

$Var(T) = \sigma_T^2 = n_1 n_2 (n_1 + n_2 + 1)/12 = 581.25, \ z = \dfrac{242 - 232.5}{\sqrt{581.25}} = .394$

p-value = 2[1-F$_Z$(.39)] = 2[1- .6517] = .6966

Therefore, reject H_0 at levels in excess of 69.66%

14.40 H_0: no difference between faculty and students for salary of the football coach

H_1: students would propose a higher salary than would faculty (one-tailed)

Sum of ranks of faculty = 2024

$T = 2042, n_1 = 50, n_2 = 50, \ E(T) = \mu_T = n_1(n_1 + n_2 + 1)/2 = 2525$

$Var(T) = \sigma_T^2 = n_1 n_2 (n_1 + n_2 + 1)/12 = 21041.667,$

$z = \dfrac{2024 - 2525}{\sqrt{21041.6667}} = -3.454$

p-value = [1-F$_Z$(3.45)] = [1- .9997] = .0003

Therefore, reject H_0 at levels in excess of 0.03%

14.42 H_0: salaries of MBA's from the two schools are equal

H_1: otherwise (two-tailed test)

$T = 1{,}243, n_1 = 30, n_2 = 30$, $E(T) = \mu_T = n_1(n_1 + n_2 + 1)/2 = 915$

$Var(T) = \sigma_T^2 = n_1 n_2 (n_1 + n_2 + 1)/12 = 4{,}575$

$$z = \frac{1{,}243 - 915}{\sqrt{4{,}575}} = 4.85, \text{ p-value} = 0.0000$$

Therefore, reject H_0 at any common level of alpha

14.44 a. Obtain rankings of the two variables

RankReturn	RankAssets
20	12
19	3
18	20
17	9
16	19
15	11
14	2
13	13
12	6
11	18
10	17
9	14
8	8
7	7
6	5
5	15
4	10
3	1
2	4
1	16

Correlations: RankReturn, RankAssets

Pearson correlation of RankReturn and RankAssets = 0.183

b. H_0: No association between rate of return and asset level

H_1: otherwise (two-tailed test)

$n = 20, r_{s.050} = .377$

Therefore, do not reject H_0 at the .10 level (two-tailed test)

c. An advantage of nonparametric tests is that normality of the variables is not assumed. The asset level variable is highly skewed and not likely to be normal. In addition, the test is less influenced by outliers and hence the extreme values of asset levels would have less weight placed on them.

14.46 H_0: No association exists between product defects and factory

H_1: otherwise

Chi-Square Test: A, B, C

Expected counts are printed below observed counts

	A	B	C	Total
1	15	25	23	63
	17.10	19.50	26.40	
2	10	12	21	43
	11.67	13.31	18.02	
3	32	28	44	104
	28.23	32.19	43.58	
Total	57	65	88	210

Chi-Sq = 0.258 + 1.551 + 0.438 + 0.239 + 0.129 + 0.493 + 0.504 + 0.546 + 0.004 = 4.162
DF = 4, P-Value = 0.385

$\chi^2_{(4,.10)} = 7.78$ Therefore, do not reject H_0 at the 10% level

14.48 H_0: No association exists between preferences of camping lanterns and job classification

H_1: otherwise

Chi-Square Test: BigStar, LoneStar, BrightStar

Expected counts are printed below observed counts

	BigStar	LoneStar	BrightStar	Total
1	54	67	39	160
	55.49	50.55	53.97	
2	23	13	44	80
	27.74	25.27	26.98	
3	69	53	59	181
	62.77	57.18	61.05	
Total	146	133	142	421

Chi-Sq = 0.040 + 5.356 + 4.151 + 0.811 + 5.960 + 10.731 + 0.618 + 0.306 + 0.069 = 28.042
DF = 4, P-Value = 0.000

$\chi^2_{(4,.005)} = 14.86$ Therefore, reject H_0 at the .5% level

14.50 H_0: No association exists between product purchase and gender

H_1: otherwise

Chi-Square Test: Male, Female

Expected counts are printed below observed counts

	Male	Female	Total
1	150	150	300
	100.00	200.00	
2	50	250	300
	100.00	200.00	
Total	200	400	600

Chi-Sq = 25.000 + 12.500 + 25.000 + 12.500 = 75.000
DF = 1, P-Value = 0.000

$\chi^2_{(1,.005)} = 7.88$ Therefore, reject H_0 at the .5% level

14.52 H_0: No association exists between family size and size of washing machine

H_1: otherwise

Chi-Square Test: 8lb, 10lb, 12lb

Expected counts are printed below observed counts

	8lb	10lb	12lb	Total
1	25	10	5	40
	9.33	16.67	14.00	
2	37	62	41	140
	32.67	58.33	49.00	
3	8	53	59	120
	28.00	50.00	42.00	
Total	70	125	105	300

Chi-Sq = 26.298 + 2.667 + 5.786 + 0.575 + 0.230 + 1.306 + 14.286 + 0.180 + 6.881 = 58.208

DF = 4, P-Value = 0.000

$\chi^2_{(4,.005)} = 14.86$ Therefore, reject H_0 at the .5% level

14.54 H_0: No association exists between location vs. product purchased

H_1: otherwise

Chi-Square Test: Tools, Lumber, Paint

Expected counts are printed below observed counts

	Tools	Lumber	Paint	Total
1	100	50	50	200
	71.67	71.67	56.67	
2	50	95	45	190
	68.08	68.08	53.83	
3	65	70	75	210
	75.25	75.25	59.50	
Total	215	215	170	600

Chi-Sq = 11.202 + 6.550 + 0.784 + 4.803 + 10.641 + 1.449 + 1.396 + 0.366 + 4.038 = 41.230

DF = 4, P-Value = 0.000

$\chi^2_{(4,.005)} = 14.86$ Therefore, reject H_0 at the .5% level

14.56 H_0: No association exists between age and use of travel agents

H_1: otherwise

Chi-Square Test: Yes, No

Expected counts are printed below observed counts

	Yes	No	Total
1	15	30	45
	21.14	23.86	
2	20	42	62
	29.12	32.88	
3	47	42	89
	41.81	47.19	
4	36	50	86
	40.40	45.60	
5	45	20	65
	30.53	34.47	
Total	163	184	347

Chi-Sq = 1.783 + 1.579 + 2.858 + 2.532 + 0.645 + 0.571 + 0.479 + 0.424 + 6.855 + 6.072 = 23.798

DF = 4, P-Value = 0.000

$\chi^2_{(4,.005)} = 14.86$ Therefore, reject H_0 at the .5% level

14.58 H_0: No association exists between management's opinion on the posting fee and the use of the internet career site

H_1: otherwise

Chi-Square Test: Yes, No

Expected counts are printed below observed counts

	Yes	No	Total
1	36	50	86
	51.78	34.22	
2	82	28	110
	66.22	43.78	
Total	118	78	196

Chi-Sq = 4.807 + 7.272 + 3.758 + 5.685 = 21.521

DF = 1, P-Value = 0.000

$\chi^2_{(1,.005)} = 7.88$ Therefore, reject H_0 at the .5% level

14.60 H_0: No association exists between provider of banking services and type of service provided

H_1: otherwise

Chi-Square Test: Bank, Retail, Other

Expected counts are printed below observed counts

	Bank	Retail	Other	Total
1	100	45	10	155
	77.50	28.84	48.66	
2	85	25	45	155
	77.50	28.84	48.66	
3	30	10	80	120
	60.00	22.33	37.67	
Total	215	80	135	430

Chi-Sq = 6.532 + 9.059 + 30.718 + 0.726 + 0.511 + 0.276 + 15.000 + 6.805 + 47.551 = 117.177

DF = 4, P-Value = 0.000

$\chi^2_{(4,.005)} = 14.86$ Therefore, reject H_0 at the .5% level

14.62 H_0: No association exists between number of dot.com company layoffs and months of service by laid-off employees

H_1: otherwise

Chi-Square Test: A, B, C

Expected counts are printed below observed counts

	A	B	C	Total
1	23	40	12	75
	25.00	35.00	15.00	
2	15	21	12	48
	16.00	22.40	9.60	
3	12	9	6	27
	9.00	12.60	5.40	
Total	50	70	30	150

Chi-Sq = 0.160 + 0.714 + 0.600 + 0.063 + 0.087 + 0.600 + 1.000 + 1.029 + 0.067 = 4.320

DF = 4, P-Value = 0.364

$\chi^2_{(4,.10)} = 7.78$ Therefore, do not reject H_0 at the 10% level

14.64 H_0: No association exists between satisfaction with an internet service and type of hardware used

H_1: otherwise

Chi-Square Test: Yes_64, No_64

Expected counts are printed below observed counts

	Yes_41	No_41	Total
1	128	40	168
	129.62	38.38	
2	45	15	60
	46.29	13.71	
3	30	8	38
	29.32	8.68	
4	30	6	36
	27.77	8.23	
Total	233	69	302

Chi-Sq = 0.020 + 0.068 + 0.036 + 0.122 + 0.016 + 0.054 + 0.178 + 0.602 = 1.096

DF = 3, P-Value = 0.778

$\chi^2_{(3,.10)} = 6.25$ Therefore, do not reject H_0 at the 10% level

14.66 a. Relation between 'easy to find' and class standing:

Using Minitab and only including those students with an opinion:

Tabulated statistics: Class, Easy to Find

```
Rows: Class    Columns: Easy to Find
                   1        2      All
1                 92       44      136
               90.94    45.06   136.00

2                 62       32       94
               62.85    31.15    94.00

3                 42       21       63
               42.13    20.87    63.00

4                 28       14       42
               28.08    13.92    42.00
Missing            0        1        *
                   *        *        *
All              224      111      335
              224.00   111.00   335.00
Cell Contents:          Count
                        Expected count
Pearson Chi-Square = 0.074, DF = 3, P-Value = 0.995
Likelihood Ratio Chi-Square = 0.074, DF = 3, P-Value = 0.995
```

b. No evidence of a significant relationship between students' class standing and their opinion on whether a student can easily find books in the college library

c. The library should test further whether all classes find it "easy" or all classes find it "difficult".

14.68 H_0: No association exists between economic status of community and use of computers

H_1: otherwise

Chi-Square Test: Poor, Affluent

```
Expected counts are printed below observed counts
          Poor  Affluent  Total
    1       75        40    115
         53.67     61.33
    2       30        80    110
         51.33     58.67
Total      105       120    225
Chi-Sq =  8.480 +  7.420 + 8.866 +  7.758 = 32.524
DF = 1, P-Value = 0.000
```

$\chi^2_{(1,.005)} = 7.88$ Reject H_0 at the .5% level, evidence that Becker's conclusions are correct.

14.70 H_0: No association exists between candy bar preference and gender

H_1: An association exists

The table below shows the breakdown of the data.

Observed Frequencies			
	Column variable		
Row variable	Male	Female	Total
Mr. Goodbar	40	10	50
Hershey's Milk Chocolate	23	70	93
Hershey's Special Dark	36	9	45
Krackel	8	4	12
Total	107	93	200

Minitab results:

Chi-Square Test: Male, Female

```
Expected counts are printed below observed counts
Chi-Square contributions are printed below expected counts
         Male  Female  Total
    1      40      10     50
        26.75   23.25
        6.563   7.551
    2      23      70     93
        49.76   43.24
       14.387  16.553
    3      36       9     45
        24.07   20.93
        5.907   6.796
    4       8       4     12
         6.42    5.58
        0.389   0.447
Total     107      93    200
Chi-Sq = 58.593, DF = 3, P-Value = 0.000
```

The hypothesis test results in p-value < 0.001. Therefore, the null hypothesis of no association is very clearly rejected, even at the 0.5% level. We conclude that there is an association between candy bar preference and gender.

14.72 Many examples of economic and business data tend to be highly skewed with extreme outliers – e.g., personal income, wealth, sales.

14.74 $H_0 : P = 0.50$ (the yen is a poor investment)

$H_1 : P \neq 0.50$ (the yen is an excellent investment)

n = 13. For 8 "excellent" and a one-sided test,

P(5 ≥ X ≥ 8) = 2P(X ≤ 5) = 2[.0001 + .0016 + .0095 + .0349 + .0873+.1571]=.581

Therefore, reject H_0 at levels of alpha in excess of 58.1%

14.76 $H_0 : P = 0.50$ (more professors believe analytical skills have deteriorated)

$H_1 : P > 0.50$ (more professors believe that analytical skills have improved)

$n = 120 - 37 = 83,\ \hat{p} = 48/83 = .5783$

$\mu = nP = 83(.5) = 41.5 \quad \sigma = .5\sqrt{n} = .5\sqrt{83} = 4.5552$

$Z = \frac{S^* - \mu}{\sigma} = \frac{47.5 - 41.5}{4.5552} = 1.32$, p-value= 1 – F_Z(1.32) = 1 - .9066 = .0934

Therefore, reject H_0 at levels of alpha in excess of 9.34%

14.78 H_0: no preference for the management candidates

H_1: otherwise (two-tailed test)

$n = 6,\quad T = \min(T_+, T_-) = 8,\quad T_{0.10} = 4$. From Appendix Table 10, $T_{0.10} = 4$

Therefore, do not reject H_0 at the 20% level

Chapter 15:

Analysis of Variance

15.2 Given the Analysis of Variance table, compute mean squares for between and for within groups. Compute the F ratio and test the hypothesis that the group means are equal.

$H_0 : \mu_1 = \mu_2 = \mu_3 = \mu_4, H_1 : otherwise$

$$MSG = \frac{SSG}{k-1}, MSW = \frac{SSW}{n-k}, F = \frac{MSG}{MSW}$$

$$MSG = \frac{879}{3}, MSW = \frac{798}{16}, F = \frac{293}{49.875} = 5.875$$

$F_{3,16,.05} = 3.24, F_{3,16,.01} = 5.29,$

Therefore, reject H_0 at the 1% level.

15.4 a. $\bar{x}_1 = 62, \bar{x}_2 = 53, \bar{x}_3 = 52$

n=16, SSW = 1028 + 1044 + 1536 = 3608

SSG = $7(62 - 56.0625)^2 + 6(53 - 56.0625)^2 + 6(52 - 56.0625)^2 = 340.9375$

SST = 3948.9375

b. Complete the anova table

One-way ANOVA: SodaSales versus CanColor

```
Analysis of Variance for SodaSale
Source     DF        SS       MS       F       P
CanColor    2       341      170    0.61   0.556
Error      13      3608      278
Total      15      3949
                                  Individual 95% CIs For Mean
                                  Based on Pooled StDev
Level       N      Mean    StDev  -+---------+---------+---------+-----
1           6     62.00    14.34            (------------*-----------)
2           5     53.00    16.16    (-------------*-------------)
3           5     52.00    19.60   (-------------*-------------)
                                  -+---------+---------+---------+-----
Pooled StDev =    16.66           36        48        60        72
```

$H_0 : \mu_1 = \mu_2 = \mu_3, H_1 : otherwise$

$F_{2,13,.05} = 3.81$, do not reject H_0 at the 5% level.

15.6 a. MINITAB Output Display:

One-way Analysis of Variance

```
Analysis of Variance
Source     DF        SS       MS        F       P
Factor      2     354.1    177.1    10.45   0.001
Error      15     254.2     16.9
Total      17     608.3
                                  Individual 95% CIs For Mean
                                  Based on Pooled StDev
Level       N      Mean    StDev  ---------+---------+---------+------
Supplier    6    32.000    3.347              (------*------)
Supplier    6    24.333    5.007   (------*------)
Supplier    6    34.833    3.817                   (------*-----)
                                  ---------+---------+---------+------
Pooled StDev =    4.116                  25.0      30.0      35.0
```

b. Assume $\alpha = 0.05$. Reject H_0 and conclude that the population mean numbers of parts per shipments not conforming to standards are not the same for all three suppliers.

c. Assume $\alpha = 0.05$.

$$MSD(k) = q\frac{S_p}{\sqrt{n}} \text{ with } S_p = \sqrt{MSW}$$

$$MSD(3) = 3.67\frac{\sqrt{16.9}}{\sqrt{18}} = 3.56$$

The Supplier B mean is significantly different from both Supplier A and Supplier C, but the later two are not different.

15.8 a. MINITAB Output Display:

One-way Analysis of Variance

```
Analysis of Variance
Source     DF        SS        MS         F        P
Factor      2        89        44      0.28    0.756
Error      18      2813       156
Total      20      2901
                                    Individual 95% CIs For Mean
                                    Based on Pooled StDev
Level       N      Mean     StDev   --+---------+---------+---------+---
Freshman    7     71.71     13.16   (-------------*--------------)
Sophomor    7     75.29     11.19        (--------------*-------------)
Juniors     7     76.57     13.05          (-------------*-------------)
                                    --+---------+---------+---------+---
Pooled StDev =    12.50            63.0      70.0      77.0      84.0
```

b. Assume $\alpha = 0.05$. Fail to reject H_0 and conclude that there is insufficient evidence that the three population mean scores are not equal.

c. Assume $\alpha = 0.05$.

$$MSD(k) = q\frac{S_p}{\sqrt{n}} \text{ with } S_p = \sqrt{MSW}$$

$$MSD(3) = 3.61\frac{\sqrt{156}}{\sqrt{21}} = 9.84$$

None of the subgroup means are significantly different from each other.

15.10 a. $\bar{x}_1 = 11.3333, \bar{x}_2 = 12.5, \bar{x}_3 = 8$, set out the anova table

One-way ANOVA: Time versus Rank

```
Analysis of Variance for Time
Source     DF        SS       MS        F        P
Rank        2     51.40    25.70     3.27    0.074
Error      12     94.33     7.86
Total      14    145.73
                                    Individual 95% CIs For Mean
                                    Based on Pooled StDev
Level       N      Mean    StDev   ---+---------+---------+---------+---
1           6    11.333    3.011             (--------*-------)
2           4    12.500    3.317               (----------*---------)
3           5     8.000    2.000   (--------*--------)
                                   ---+---------+---------+---------+---
Pooled StDev =    2.804               6.0       9.0      12.0      15.0
```

b. $H_0: \mu_1 = \mu_2 = \mu_3, H_1: otherwise$

$F_{2,12,.05} = 3.89$, do not reject H_0 at the 5% level.

15.12 MINITAB Output Display:

One-way Analysis of Variance

```
Analysis of Variance
Source      DF        SS        MS        F        P
Factor       2     48.96     24.48     4.07    0.039
Error       15     90.13      6.01
Total       17    139.09
                                     Individual 95% CIs For Mean
                                     Based on Pooled StDev
Level        N      Mean     StDev   --+---------+---------+---------+---
Confessi     6    10.402     2.268                    (--------*-------)
People W     6     7.045     2.182     (-------*--------)
Newsweek     6     6.777     2.850    (-------*--------)
                                     --+---------+---------+---------+---
Pooled StDev =     2.451             5.0       7.5      10.0      12.5
```

Assume $\alpha = 0.05$. Reject H_0 and conclude that the population mean fog indices are not the same for all three magazines.

$$MSD(k) = q\frac{S_p}{\sqrt{n}} \text{ with } S_p = \sqrt{MSW}$$

$$MSD(3) = 3.61\frac{\sqrt{6.01}}{\sqrt{18}} = 2.09$$

The True Confessions mean is significantly different from both People Weekly and Newsweek, but the later two are not different.

15.14 a. $\hat{\mu} = 8.0744$

b. $\hat{G}_1 = 10.4017 - 8.0744 = 2.3273$, $\hat{G}_2 = 7.045 - 8.0744 = -1.0294$

$\hat{G}_3 = 6.7767 - 8.0744 = -1.2977$

c. $\hat{\varepsilon}_{32} = 11.15 - 10.4017 = .7483$

15.16 $H_0: \mu_1 = \mu_2 = \mu_3 = \mu_4, H_1: otherwise$

$R_1 = 49, \quad R_2 = 84, \quad R_3 = 76, \quad R_4 = 81$

$$W = \frac{12}{23(24)}\left[(49^2/4) + (84^2/6) + (76^2/7) + (81^2/6)\right] - 3(24) = 8.32$$

$\chi^2_{(3,.05)} = 7.81$, therefore, reject H_0 at the 5% level

15.18 $H_0: \mu_1 = \mu_2 = \mu_3, H_1: otherwise$

$R_1 = 61, \quad R_2 = 37, \quad R_3 = 38$

$$W = \frac{12}{16(17)}\left[(3721/6) + (1396/5) + (1444/5)\right] - 3(17) = 1.18$$

$\chi^2_{(2,.10)} = 4.61$, therefore, do not reject H_0 at the 10% level

Using Minitab:

Kruskal-Wallis Test: SodaSales versus CanColor

```
Kruskal-Wallis Test on SodaSale
CanColor     N    Median    Ave Rank        Z
1            6     60.00        10.2     1.08
2            5     52.00         7.4    -0.62
3            5     53.00         7.6    -0.51
Overall     16                   8.5
H = 1.18  DF = 2  P = 0.554
H = 1.19  DF = 2  P = 0.553 (adjusted for ties)
```

15.20 $H_0: \mu_1 = \mu_2 = \mu_3, H_1: otherwise$

$R_1 = 63.5, \quad R_2 = 26, \quad R_3 = 81.5$

$$W = \frac{12}{18(19)}\left[(4032.25 + 676 + 6642.25)/6\right] - 3(19) = 9.3772$$

$\chi^2_{(2,.01)} = 9.21$, therefore, reject H_0 at the 1% level

Using Minitab:

Kruskal-Wallis Test: Nonconforming versus Supplier

```
Kruskal-Wallis Test on Nonconfo
Supplier     N    Median    Ave Rank        Z
1            6     32.00        10.6     0.61
2            6     24.50         4.3    -2.90
3            6     35.00        13.6     2.29
Overall     18                   9.5
H = 9.38  DF = 2  P = 0.009
H = 9.47  DF = 2  P = 0.009 (adjusted for ties)
```

15.22 $H_0: \mu_1 = \mu_2 = \mu_3, H_1: otherwise$

$R_1 = 66, \quad R_2 = 79.5, \quad R_3 = 85.5$

$$W = \frac{12}{21(22)}\left[(4356 + 6320.25 + 7310.25)/7\right] - 3(22) = .7403$$

$\chi^2_{(2,.10)} = 4.61$, therefore, do not reject H_0 at the 10% level

15.24 $H_0: \mu_1 = \mu_2 = \mu_3, H_1: otherwise$

$R_1 = 54.5, \quad R_2 = 43.5, \quad R_3 = 22$

$$W = \frac{12}{15(16)}\left[(2970.25/6) + (1892.25/4) + (484/5)\right] - 3(16) = 5.2452$$

$\chi^2_{(2,.10)} = 4.61$, therefore, reject H_0 at the 10% level

15.26 a. The null hypothesis tests the equality of the population mean ratingsacross the classes

b. $H_0: \mu_1 = \mu_2 = \mu_3, H_1: otherwise$

$W = .17$, $\chi^2_{(2,.10)} = 4.61$, therefore, do not reject H_0 at the 10% level

15.28 $MSG = \frac{SSG}{K-1}$, $MSB = \frac{SSB}{H-1}$, $MSE = \frac{SSE}{(K-1)(H-1)}$

Test of K population group means all the same:

Reject Ho if $\frac{MSG}{MSE} > F_{K-1,(K-1)(H-1),\alpha}$

Test of H population block means all the same:

Reject Ho if $\frac{MSB}{MSE} > F_{H-1,(K-1)(H-1),\alpha}$

$MSG = \frac{380}{6} = 63.333$, $MSB = \frac{232}{5} = 46.4$, $MSE = \frac{387}{(6)(5)} = 12.90$

$\frac{MSB}{MSE} = \frac{46.4}{12.90} = 3.597$, $F_{5,30,.05} = 2.53$, $F_{5,30,.01} = 3.70$

Reject at the 5% level, do not reject H_0 at the 1% level that the block means differ.

$\frac{MSG}{MSE} = \frac{63.333}{12.90} = 4.91$, $F_{6,30,.05} = 2.42$, $F_{6,30,.01} = 3.47$

Reject H_0 at the 1% level. Evidence suggests the group means differ

15.30 a. **Two-way ANOVA: earngrowth versus OilCo, Analyst**

```
Analysis of Variance for earngrow
Source         DF         SS         MS         F        P
OilCo           4       3.30       0.83      0.31    0.866
Analyst         3      31.35      10.45      3.93    0.036
Error          12      31.90       2.66
Total          19      66.55
                                Individual 95% CI
OilCo            Mean   ------+---------+---------+---------+-----
1               10.00             (-------------*--------------)
2                9.50   (--------------*--------------)
3               10.25             (-------------*--------------)
4               10.75                 (--------------*-------------)
5               10.25             (-------------*--------------)
                        ------+---------+---------+---------+-----
                             8.40      9.60     10.80     12.00
                                Individual 95% CI
Analyst_         Mean   -----+---------+---------+---------+------
A                9.80        (---------*---------)
B                9.80        (---------*---------)
C                8.80   (---------*---------)
D               12.20                     (---------*---------)
                        -----+---------+---------+---------+------
                           8.00      9.60     11.20     12.80
```

b. $H_0: \mu_1 = \mu_2 = \mu_3 = \mu_4 = \mu_5, H_1: Otherwise$ $F_{4,12,.05} = 3.26 > .31$, therefore, do not reject H_0 at the 5% level

15.32 a. two-way ANOVA table:

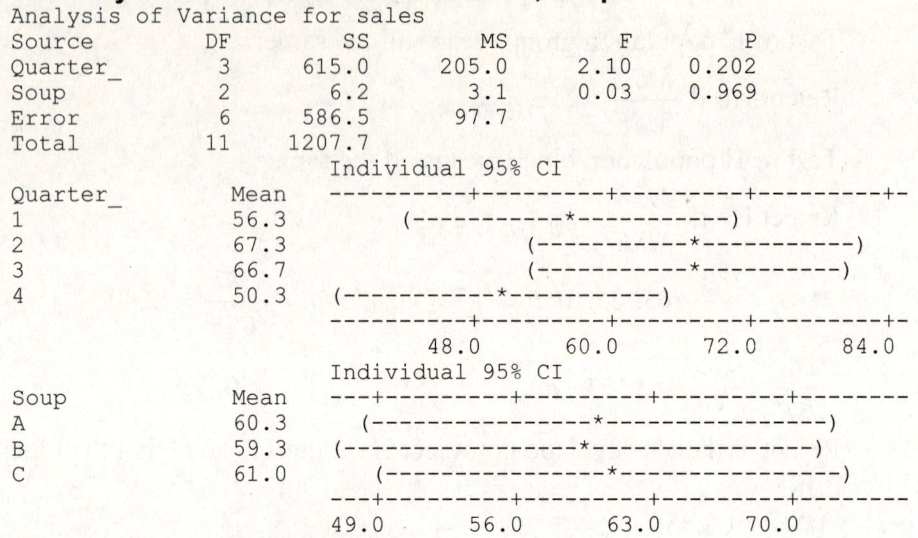

Two-way ANOVA: sales versus Quarter, soup

```
Analysis of Variance for sales
Source        DF        SS         MS        F        P
Quarter_       3     615.0      205.0     2.10    0.202
Soup           2       6.2        3.1     0.03    0.969
Error          6     586.5       97.7
Total         11    1207.7
                            Individual 95% CI
Quarter_         Mean   ----------+---------+---------+---------+-
1                56.3        (-----------*-----------)
2                67.3                 (-----------*-----------)
3                66.7                 (-----------*----------)
4                50.3   (-----------*-----------)
                        ----------+---------+---------+---------+-
                               48.0      60.0      72.0      84.0
                            Individual 95% CI
Soup             Mean   ---+---------+---------+---------+--------
A                60.3     (----------------*----------------)
B                59.3   (----------------*----------------)
C                61.0      (----------------*----------------)
                        ---+---------+---------+---------+--------
                        49.0      56.0      63.0      70.0
```

b. $H_0 : \mu_1 = \mu_2 = \mu_3, H_1 : otherwise$

$F_{2,6,.05} = 5.14 > .03$, therefore, do not reject H_0 at the 5% level

15.34 a. two-way ANOVA table:

Two-way ANOVA: Ratings versus Exam, Text

```
Analysis of Variance for Ratings
Source        DF        SS         MS        F        P
Exam           2    0.2022     0.1011     5.20    0.077
Text           2    0.4356     0.2178    11.20    0.023
Error          4    0.0778     0.0194
Total          8    0.7156
                            Individual 95% CI
Exam             Mean   ----------+---------+---------+---------+-
Essays          4.633   (-----------*----------)
MC              5.000                    (-----------*----------)
Mix             4.833            (-----------*----------)
                        ----------+---------+---------+---------+-
                              4.600     4.800     5.000     5.200
                            Individual 95% CI
Text             Mean   ---+---------+---------+---------+--------
A               4.667     (--------*--------)
B               5.133                    (--------*--------)
C               4.667     (--------*--------)
                        ---+---------+---------+---------+--------
                        4.500     4.750     5.000     5.250
```

b. $H_0 : \mu_1 = \mu_2 = \mu_3, H_1 : otherwise$ [texts]

$F_{2,4,.05} = 6.94 < 11.20$, therefore, reject H_0 at the 5% level

c. $H_0 : \mu_1 = \mu_2 = \mu_3, H_1 : otherwise$ [exam type]

$F_{2,4,.05} = 6.94 > 5.20$, therefore, do not reject H_0 at the 5% level

15.36 $\hat{G}_3 = -.1556$

$\hat{B}_1 = .1778$

$\hat{\varepsilon}_{31} = .0556$

15.38 a. complete the ANOVA table:

Source of Variation	Sum of Squares	df	Mean square	F Ratio
Fertilizers	135.6	3	45.20	6.0916
Soil Types	81.7	5	16.34	2.2022
Error	111.3	15	7.42	
Total	328.6	23		

b. $H_0 : \mu_1 = \mu_2 = \mu_3 = \mu_4, H_1 : otherwise$ [fertilizers]

$F_{3,15,.01} = 5.42 < 6.0916$, therefore, reject H_0 at the 1% level

c. $H_0 : \mu_1 = \mu_2 = \mu_3 = \mu_4 = \mu_5 = \mu_6, H_1 : Otherwise$ [soil types]

$F_{5,15,.05} = 2.90 > 2.2021$, therefore, do not reject H_0 at the 5% level

15.40 Given, say, ten pairs observations, the F statistic would have 1, 9 degrees of freedom. The test is in the form of a two-tailed test. With alpha = .05, the critical value of F would be 5.12. For a matched – pairs test, the degrees of freedom would be 9, and the area in each tail would be .025. The critical value for t would be 2.262 (which is the square root of the F statistic of 5.12). Therefore, the two tests are equivalent.

15.42 $MSG = \frac{SSG}{K-1},\ MSB = \frac{SSB}{H-1},\ MSI = \frac{SSI}{(K-1)(H-1)},\ MSE = \frac{SSE}{KH(L-1)},\ F = \frac{MSG}{MSE}, \frac{MSB}{MSE}, \frac{MSI}{MSE}$

$$MSG_A = \frac{86}{4} = 21.5,\ MSG_B = \frac{75}{5} = 15,\ MSI = \frac{75}{20} = 3.75,\ MSE = \frac{300}{90} = 3.33$$

$$F\ Ratio : Interaction = \frac{MSI}{MSE} = \frac{3.75}{3.33} = 1.125,\ F_{30,80,.05} \approx 1.65,\ F_{30,80,.01} \approx 2.03$$

Do not reject H_0 at the 5% level. No significant interaction exists between treatment groups A and B. Therefore, go on to test the main effects of each treatment group.

$$F\ Ratio : Treatment\ A = \frac{MSG_A}{MSE} = \frac{21.5}{3.33} = 6.456,\ F_{4,80,.05} \approx 2.53,\ F_{4,80,.01} \approx 3.65$$

Reject H_0 at the 1% level, there is a significant main effect for group A.

$$F\ Ratio : Treatment\ B = \frac{MSG_B}{MSE} = \frac{15}{3.33} = 4.505,\ F_{5,80,.05} \approx 2.37,\ F_{5,80,.01} \approx 3.34$$

Reject H_0 at the 1% level, there is a significant main effect for group B.

15.44 a. ANOVA table:

Source of Variation	Sum of Squares	df	Mean square	F Ratio
Contestant	364.50	21	17.3571	19.2724
Judges	.81	8	.1013	.1124
Interaction	4.94	168	.0294	.0326
Error	1069.94	1188	.9006	
Total	1440.19	1385		

H_0: Mean value for all 22 contestants is the same
H_1: Otherwise
$F_{21,1188,.01} \approx 1.88 < 19.2724$, therefore, reject H_0 at the 1% level

H_0: Mean value for all 9 judges is the same
H_1: Otherwise
$F_{8,1188,.05} \approx 1.94 > .1124$, therefore, do not reject H_0 at the 5% level

H_0: No interaction exists between contestants and judges
H_1: Otherwise
$F_{168,1188,.05} \approx 1.22 > .0326$, therefore, do not reject H_0 at the 5% level

15.46 a. ANOVA table:

Source of Variation	Sum of Squares	df	Mean square	F Ratio
Test type	57.5556	2	28.7778	4.7091
Subject	389.0000	3	129.6667	21.2182
Interaction	586.0000	6	97.66667	15.9818
Error	146.6667	24	6.1111	
Total	1179.2223	35		

b. H_0: No interaction exists between contestants and judges
H_1: Otherwise
$F_{6,24,.01} = 3.67 < 15.9818$, therefore, reject H_0 at the 1% level

15.48 a. The implied assumption is that there is no interaction effect between student year and dormitory ratings
b. Using Minitab:

General Linear Model: Ratings_48 versus Dorm_48, Year_48

```
Factor      Type Levels Values
Dorm_48    fixed      4 A B C D
Year_48    fixed      4 1 2 3 4

Analysis of Variance for Ratings_, using Adjusted SS for Tests
Source       DF      Seq SS      Adj SS      Adj MS        F      P
Dorm_48       3      20.344      20.344       6.781     4.91  0.008
Year_48       3      10.594      10.594       3.531     2.56  0.078
Error        25      34.531      34.531       1.381
Total        31      65.469
```

Source of Variation	Sum of Squares	df	Mean square	F Ratio
Dorm	20.344	3	6.781	4.91
Year	10.594	3	3.531	2.56
Error	34.531	25	1.381	
Total	65.469	31		

c. H_0: Mean ratings for all 4 dormitories is the same
   H_1: Otherwise
   $F_{3,25,.01} = 4.68 < 4.91$, therefore, reject H_0 at the 1% level

d. H_0: Mean ratings for all 4 student years is the same
   H_1: Otherwise
   $F_{3,25,.05} = 2.99 > 2.56$, therefore, do not reject H_0 at the 5% level

15.50

Source of Variation	Sum of Squares	df	Mean square	F Ratio
Color	243.250	2	121.625	11.3140
Region	354.000	3	118.000	10.9767
Interaction	189.750	6	31.625	2.9419
Error	129.000	12	10.750	
Total	916.000	23		

H_0: No interaction exists between region and can color,
H_1: Otherwise
$F_{6,12,.01} = 4.82 > 2.9419$, therefore, do not reject H_0 at the 1% level
Main effects are significant for both Color and Region

15.52 One-way ANOVA examines the effect of a single factor (having three or more conditions). Two-way ANOVA recognizes situations in which more than one factor may be significant.

15.54

Source of Variation	Sum of Squares	df	Mean square	F Ratio
Between	5156	2	2578.000	21.4458
Within	120802	1005	120.201	
Total	125967	1007		

$F_{2,1005,.01} = 4.61 < 21.4458$, therefore, reject H_0 at the 1% level

15.56 a. Use result from Equation 15.2 to find SSG, then compute MSG and finally, find SSW using the fact that SSW = $(n-k)\dfrac{MSG}{F}$

Source of Variation	Sum of Squares	df	Mean square	F Ratio
Between	221.3400	3	73.7800	25.6
Within	374.6640	130	2.8820	
Total	596.0040	133		

b. H_0: Mean salaries are the same for managers in all 4 groups
H_1: Otherwise
$F_{3,130,.01} \approx 3.95 < 25.6$, therefore, reject H_0 at the 1% level

15.58

Source of Variation	Sum of Squares	df	Mean square	F Ratio
Between	11438.3028	2	5719.1514	.7856
Within	109200.000	15	7280.000	
Total	120638.3028	17		

H_0: Mean sales levels are the same for all three periods
H_1: Otherwise
$F_{2,15,.05} = 3.68 > .7856$, therefore, do not reject H_0 at the 5% level

15.60 $H_0: \mu_1 = \mu_2 = \mu_3 = \mu_4, H_1: otherwise$

$R_1 = 48.5, \quad R_2 = 55, \quad R_3 = 74, \quad R_4 = 32.5$

$$W = \frac{12}{20(21)}[(2352.25+3025+5476+1056.25)/5]-3(21) = 5.0543$$

$\chi^2_{(3,.10)} = 6.25$, therefore, do not reject H_0 at the 10% level

15.62 a. $$SSW = \sum_{j=1}^{K}\sum_{i=1}^{n_i}(x_{ij}-\bar{x}_i)^2$$

$$= \sum_{j=1}^{K}\left[\sum_{i=1}^{n_i} x^2_{ij} - 2n_i\bar{x}^2_i + n_i\bar{x}^2_i\right]$$

$$= \sum_{j=1}^{K}\sum_{i=1}^{n_i} x^2_{ij} - \sum_{i=1}^{K} n_i\bar{x}^2_i$$

b. $SSG = \sum_{j=1}^{K} n_i(\overline{x}_i - \overline{x})^2$

$$= \sum_{i=1}^{K} n_i \overline{x}^2{}_i - 2\overline{x}\sum_{i=1}^{k} n_i \overline{x}_i + n\overline{x}^2$$

$$= \sum_{i=1}^{K} n_i \overline{x}^2{}_i - 2n\overline{x}^2 + n\overline{x}^2$$

$$= \sum_{i=1}^{K} n_i \overline{x}^2{}_i - n\overline{x}^2$$

c. $SST = \sum_{i=1}^{K}\sum_{j=1}^{n_i}(x_{ij} - \overline{x}_i)^2$

$$= \sum_{i=1}^{K}\left[\sum_{i=1}^{n_i} x^2{}_{ij} - 2\overline{x}\sum_{i=1}^{n_i} x_i + n_i\overline{x}^2{}_i\right]$$

$$= \sum_{i=1}^{K}\sum_{i=1}^{n_i}(x^2{}_{ij} - 2n_i\overline{x}\,\overline{x}_i + n_i x^2)$$

$$= \sum_{i=1}^{K}\sum_{j=1}^{n_i} x^2{}_{ij} - n\overline{x}^2$$

15.64

Source of Variation	Sum of Squares	df	Mean square	F Ratio
Consumers	37571.5	124	302.996	1.3488
Brands	32987.3	2	16493.65	73.4226
Error	55710.7	248	224.6399	
Total	126269.5	374		

H_0: Mean perception levels are the same for all three brands

H_1: Otherwise

$F_{2,248,.01} \approx 4.79 < 73.4226$, therefore, reject H_0 at the 1% level

15.66 Using Minitab:

Two-way ANOVA: GPA versus SAT, Income

```
Source      DF        SS        MS      F      P
SAT          2  0.826667  0.413333  24.80  0.006
Income       2  0.006667  0.003333   0.20  0.826
Error        4  0.066667  0.016667
Total        8  0.900000
S = 0.1291   R-Sq = 92.59%   R-Sq(adj) = 85.19%
                 Individual 95% CIs For Mean Based on
                 Pooled StDev
SAT       Mean   +---------+---------+---------+---------
High   3.36667                  (------*------)
Mod    2.90000   (------*------)
VeryH  3.63333                           (------*------)
                 +---------+---------+---------+---------
               2.70      3.00      3.30      3.60
                    Individual 95% CIs For Mean Based on
                    Pooled StDev
Income      Mean    ------+---------+---------+---------+---
High     3.33333        (-------------*-------------)
Low      3.26667    (-------------*-------------)
Mod      3.30000      (-------------*-------------)
                    ------+---------+---------+---------+---
                        3.15      3.30      3.45      3.60
```

Source of Variation	Sum of Squares	df	Mean square	F Ratio
Income	.0067	2	.0033	.2000
SAT Score	.8267	2	.4133	24.8000
Error	.0667	4	.0167	
Total	.9000	8		

H_0: Mean gpa's are the same for all three income groups
H_1: Otherwise
$F_{2,4,.05} = 6.94 > .2000$, therefore, do not reject H_0 at the 5% level

H_0: Mean gpa's are the same for all three SAT score groups
H_1: Otherwise
$F_{2,4,.01} = 18.0 < 24.8$, therefore, reject H_0 at the 1% level

Descriptive Statistics: GPA

```
Variable   Mean  SE Mean  StDev  Variance  Minimum     Q1  Median     Q3
GPA       3.300    0.112  0.335     0.112    2.800  2.950   3.400  3.600

Variable  Maximum    IQR
GPA         3.700  0.650
```

15.68 a. $\hat{\mu} = 3.3$

b. $\hat{G}_2 = 0.0$

c. $\hat{B}_2 = .0667$

d. $\hat{\varepsilon}_{22} = .1333$

15.70 ANOVA table:

Source of Variation	Sum of Squares	df	Mean square	F Ratio
Prices	.178	2	.0890	.0944
Countries	4.365	2	2.1825	2.3151
Interaction	1.262	4	.3155	.3347
Error	93.330	99	.9427	
Total	99.135	107		

H_0: Mean quality ratings for all three prices levels is the same
H_1: Otherwise
$F_{2,99,.05} \approx 3.07 > .0944$; therefore, do not reject H_0 at the 5% level

H_0: Mean quality ratings for all three countries is the same
H_1: Otherwise
$F_{2,99,.05} \approx 3.07 > 2.3151$; therefore, do not reject H_0 at the 5% level

H_0: No interaction exists between price and country
H_1: Otherwise
$F_{4,99,.05} \approx 2.45 > .3347$, therefore, do not reject H_0 at the 5% level

15.72 a. ANOVA table:

Source of Variation	Sum of Squares	df	Mean square	F Ratio
Income	.0178	2	.0089	.5333
SAT score	2.2011	2	1.1006	66.0333
Interaction	.1022	4	.0256	1.5333
Error	.1500	9	.0167	
Total	2.4711	17		

b. H_0: Mean GPAs for all three income groups is the same
H_1: Otherwise
$F_{2,9,.05} = 4.26 > .5333$, therefore, do not reject H_0 at the 5% level

c. H_0: Mean GPAs for all three SAT score groups is the same
H_1: Otherwise
$F_{2,9,.01} = 8.02 < 66.0333$; therefore, reject H_0 at the 1% level

d. H_0: No interaction exists between income and SAT score group
H_1: Otherwise
$F_{4,9,.05} = 3.63 > 1.5333$; therefore, do not reject H_0 at the 5% level

Chapter 16:

Time-Series Analysis and Forecasting

16.2 Refer to Figure 16.4 Compute the revised Laspeyres Quantity Index for years 1 through 6 if the year one prices are 1.45 (wheat), 1.21 (corn), and 2.98 (soybeans).

Year	Wheat	Corn	Soybeans	Total Cost	Laspeyres Quantity Index
	1.45	1.21	2.98		
1	1,352	4,152	1,127	10,343	100.0
2	1,618	5,641	1,176	12,676	122.6
3	1,545	5,573	1,271	12,771	123.5
4	1,705	5,647	1,547	13,915	134.5
5	2,122	5,829	1,547	14,740	142.5
6	2,142	6,266	1,288	14,526	140.4
7	2,026	6,357	1,716	15,743	152.2
8	1,799	7,082	1,843	16,670	161.2
9	2,134	7,939	2,268	19,459	188.1
10	2,370	6,648	1,817	16,895	163.4

16.4 a. e.g., $I_2 = 100(35.875/35) = 102.5$

b. e.g., $I_1 = 100(35/34.375) = 101.82$

Week	Price	BaseWeek1	BaseWeek4
1	35	100.00	101.82
2	35.875	102.50	104.36
3	34.75	99.29	101.09
4	34.375	98.21	100.00
5	35	100.00	101.82
6	34.875	99.64	101.45
7	35	100.00	101.82
8	34.75	99.29	101.09
9	34.75	99.29	101.09
10	35.25	100.71	102.55
11	38.75	110.71	112.73
12	37.125	106.07	108.00

16.6 a. Unweighted average index

Year	Average	Index of Average
1	11.8	100.00
2	12.43	105.37
3	12.93	109.60
4	13.30	112.71
5	13.63	115.54
6	13.83	117.23

b. Laspeyres index

Year	$\sum q_{oi}p_{1i}$	Laspeyres Index
1	1120.4	100.00
2	1174.30	104.81
3	1237.90	110.49
4	1256.40	112.14
5	1296.40	115.71
6	1316.10	117.47

16.8 A price index for energy is helpful in that it allows us to say something about price movements over time for a group of commodities, namely, energy prices. A weighted index of prices allows one to compare the cost of a group of products across periods.

16.10 A time series contains 50 observations, find the probability that the number of runs

a. is less than 14

$$Z = \frac{R - \frac{n}{2} - 1}{\sqrt{\frac{n^2 - 2n}{4(n-1)}}} = Z = \frac{14 - \frac{50}{2} - 1}{\sqrt{\frac{50^2 - 2(50)}{4(50-1)}}} = -3.43 \quad P(Z < -3.43) = .0003$$

b. is less than 17

$$Z = \frac{R - \frac{n}{2} - 1}{\sqrt{\frac{n^2 - 2n}{4(n-1)}}} = Z = \frac{17 - \frac{50}{2} - 1}{\sqrt{\frac{50^2 - 2(50)}{4(50-1)}}} = -2.57 \quad P(Z < -2.57) = .0051$$

c. is greater than 38

$$Z = \frac{R - \frac{n}{2} - 1}{\sqrt{\frac{n^2 - 2n}{4(n-1)}}} = Z = \frac{38 - \frac{50}{2} - 1}{\sqrt{\frac{50^2 - 2(50)}{4(50-1)}}} = 3.43 \quad P(Z > 3.43) = .0003$$

16.12 Runs test on Value of the exchange rate

Runs Test: Value

```
Value    K =      99.8500
The observed number of runs =    6
    The expected number of runs =     7.0000
     6 Observations above K     6 below
  * N Small -- The following approximation may be invalid
              The test is significant at  0.5448
              Cannot reject at alpha = 0.05
```

Do not reject Ho that the data series is random. There is no evidence of nonrandom patterns in the data.

16.14 Runs test on Stock Market Return

Runs Test: Return

```
Return   K =      17.5000
   The observed number of runs =    9
    The expected number of runs =    8.0000
     7 Observations above K     7 below
  * N Small -- The following approximation may be invalid
              The test is significant at  0.5780
```

Do not reject Ho that the data series is random. There is no evidence of nonrandom patterns in the data.

16.16 a. Runs test on Housing Starts

Runs Test: Starts

```
Starts   K =      7.3500
   The observed number of runs =  10
     The expected number of runs =  13.0000
     12 Observations above K    12 below
                The test is significant at  0.2105
                Cannot reject at alpha = 0.05
```

Do not reject Ho that the data series is random. There is no evidence of nonrandom patterns in the data.

b. From the time series plot below, strong cyclical behavior is evident.

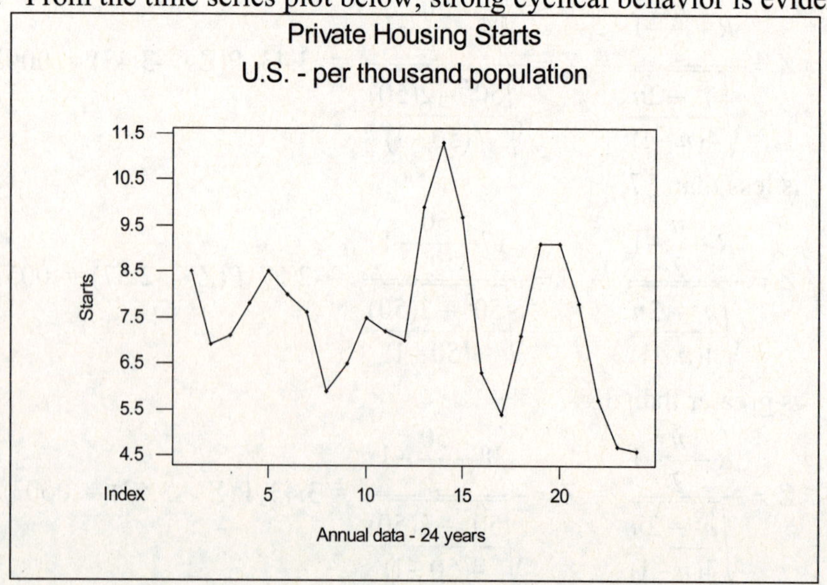

16.18 a. Time series plot of Quarterly sales
Data patterns evident in the time series plot include; strong seasonality and a strong upward trend

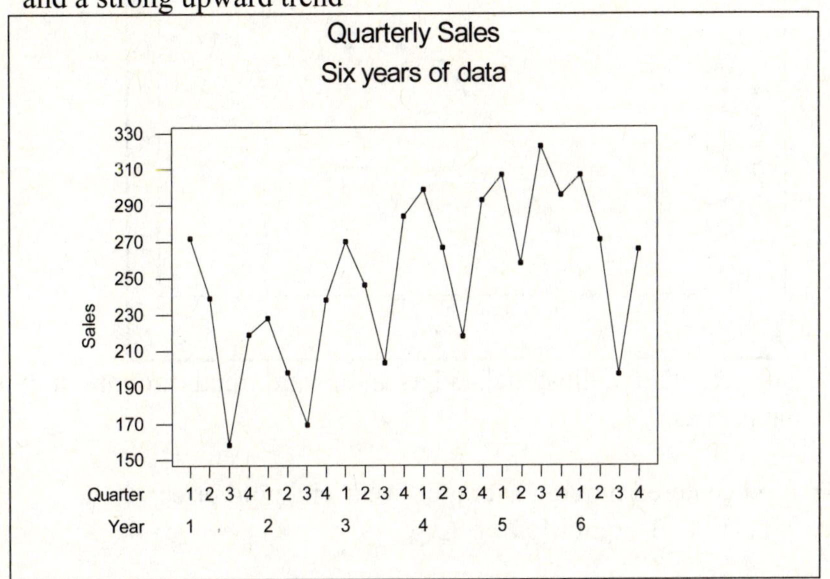

b.

Period	4 Period MA	$100\frac{X_t}{X_t^*}$	Seas. Factor	Adj. Series
1-1			112.848	241.032
2			99.609	239.938
3	216.500	72.979	79.826	197.930
4	205.875	106.375	107.716	203.312
2-1	202.125	112.802		202.041
2	205.875	96.175		198.777
3	213.500	79.157		211.709
4	224.750	105.895		220.951
3-1	235.000	114.894		239.259
2	245.000	100.408		246.966
3	254.375	79.803		254.302
4	260.625	108.969		263.656
4-1	265.125	112.777		264.958
2	268.125	99.580		268.048
3	270.250	80.666		273.093
4	270.125	108.468		272.011
5-1	270.750	113.389		272.047
2	272.875	94.549		259.013
3	273.250	84.904		290.631
4	274.875	107.685		274.796
6-1	272.125	112.816		272.047
2	264.000	102.652		272.064
3				246.786
4				246.945

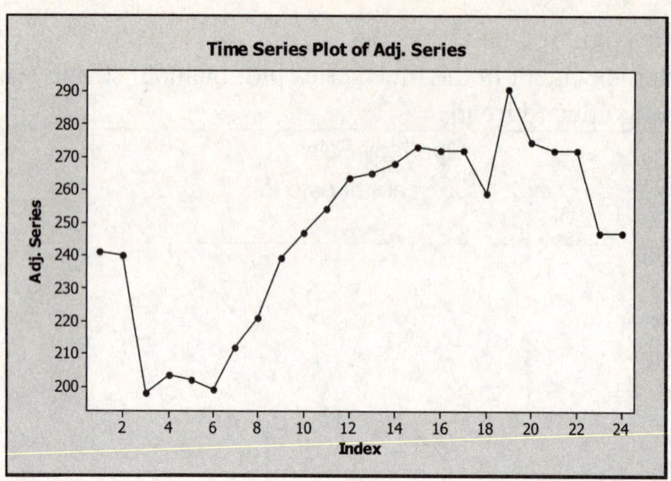

The seasonally adjusted data has an upward trend throughout much of the data series.

16.20 3-period centered moving average – year-end gold price

Year	3-point Moving Avg
1	*
2	176.000
3	308.667
4	450.333
5	507.667
6	480.000
7	411.333
8	381.000
9	340.667
10	347.333
11	406.667
12	433.667
13	421.667
14	*

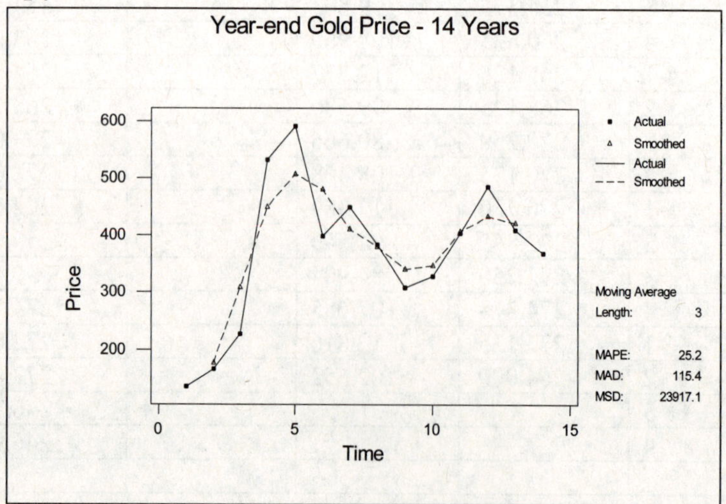

The resulting data shows strong cyclical behavior

16.22

Year	7ptMA
1	*
2	*
3	*
4	30.4429
5	25.3429
6	23.0000
7	23.4286
8	22.7429
9	21.1286
10	20.6000
11	23.1286
12	28.1000
13	32.0857
14	33.8000
15	35.1571
16	35.7143
17	34.9714
18	31.5143
19	27.4571
20	27.0857
21	34.3286
22	40.5429
23	43.7143
24	48.6000
25	54.4286
26	*
27	*
28	*

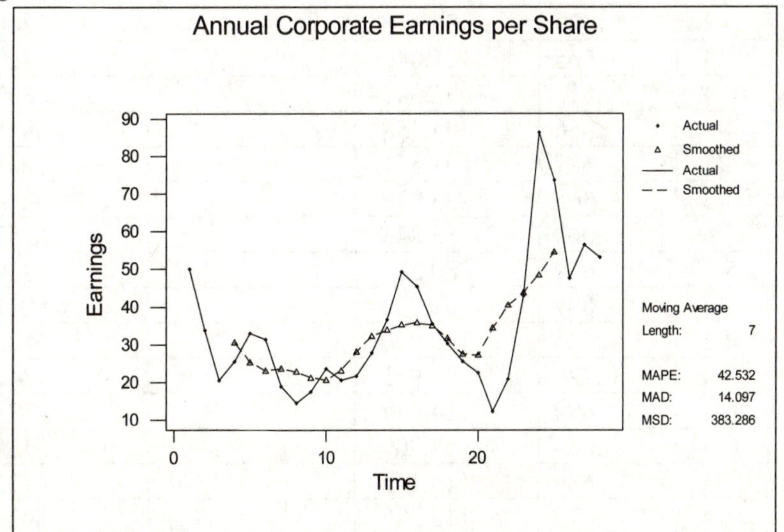

The smoothed data exhibits a cyclical data pattern

16.24 a.

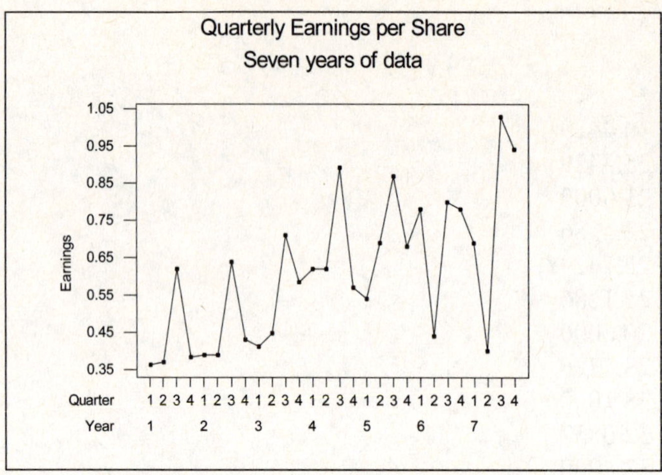

The data series exhibits a strong seasonal component.

b.

Period	4 Period MA	$100\frac{X_t}{X_t^*}$	Seas. Factor	Adj. Series
1-1			90.930	.3981
2			86.020	.4301
3	.438	141.902	130.400	.4762
4	.443	86.608	92.649	.4145
2-1	.448	86.830		.4278
2	.456	85.284		.4522
3	.465	137.493		.4900
4	.475	90.761		.4652
3-1	.491	83.643		.4520
2	.520	86.216		.5208
3	.565	126.046		.5460
4	.613	95.347		.6303
4-1	.656	94.458		.6818
2	.677	91.581		.7208
3	.665	133.935		.6833
4	.664	85.843		.6152
5-1	.670	80.582		.5939
2	.681	101.284		.8021
3	.725	120.000		.6672
4	.724	93.955		.7340
6-1	.684	114.077		.8578
2	.688	64.000		.5115
3	.689	116.153		.6135
4	.673	115.985		.8419
7-1	.696	99.102		.7588
2	.745	53.691		.4650
3				.7899
4				1.0146

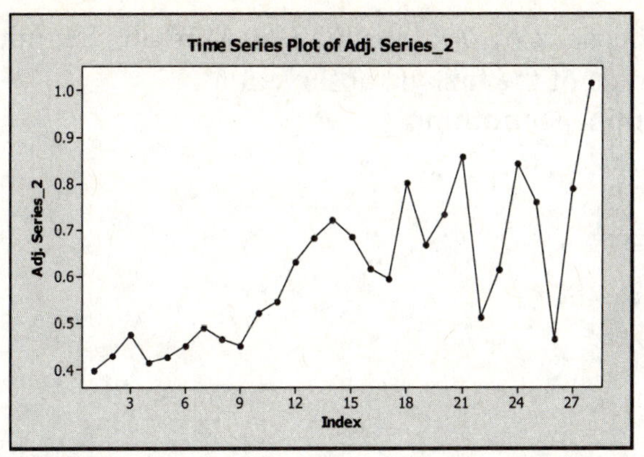

Seasonally adjusted series shows a strong upward trend.

16.26

Period	4 Period MA	$100 \frac{X_t}{X_t^*}$	Seas. Factor	Adj. Series
1-1			93.701	574.165
2			97.648	634.935
3			119.127	729.476
4			101.932	832.911
5			102.038	928.084
6			97.450	955.365
7	767.750	97.167	99.356	750.834
8	773.750	95.638	100.300	737.789
9	775.000	84.387	87.504	747.395
10	770.333	113.457	113.255	771.707
11	759.667	99.912	98.563	770.069
12	743.000	85.734	89.127	714.721
2-1	730.250	87.094		678.752
2	725.750	91.767		682.043
3	721.458	118.233		716.045
4	712.542	105.678		738.730
5	699.542	112.502		771.280
6	689.458	100.224		709.084
7	684.000	99.415		684.406
8	678.917	102.811		695.915
9	668.208	88.745		677.684
10	651.750	110.625		636.614
11	630.917	95.100		608.751
12	611.417	90.609		621.586
3-1	598.167	98.300		627.526
2	583.625	101.435		606.260
3	570.375	117.467		562.426
4	563.542	96.000		530.748
5	558.250	89.387		489.033
6	551.917	92.586		524.373
7				545.512
8				485.545
9				555.403
10				586.286
11				537.730
12				529.583

16.28 Use smoothing constant of .7 (alpha of .3) in Minitab. Set initial smoothing value at the average of the first '1' observations

Single Exponential Smoothing

```
Data           Price
Length         14.0000
NMissing       0
Smoothing Constant
Alpha: 0.7
Accuracy Measures
MAPE:     20.7
MAD:      84.3
MSD:   13759.9
```

Row	Time	Price	SMOO2	FITS2	Error
1	1	135	135.000	135.000	0.000
2	2	166	156.700	135.000	31.000
3	3	227	205.910	156.700	70.300
4	4	533	434.873	205.910	327.090
5	5	591	544.162	434.873	156.127
6	6	399	442.549	544.162	-145.162
7	7	450	447.765	442.549	7.451
8	8	385	403.829	447.765	-62.765
9	9	308	336.749	403.829	-95.829
10	10	329	331.325	336.749	-7.749
11	11	405	382.897	331.325	73.675
12	12	486	455.069	382.897	103.103
13	13	410	423.521	455.069	-45.069
14	14	369	385.356	423.521	-54.521

Row	Period	Forecast	Lower	Upper
1	15	385.356	178.884	591.828
2	16	385.356	178.884	591.828
3	17	385.356	178.884	591.828
4	18	385.356	178.884	591.828
5	19	385.356	178.884	591.828

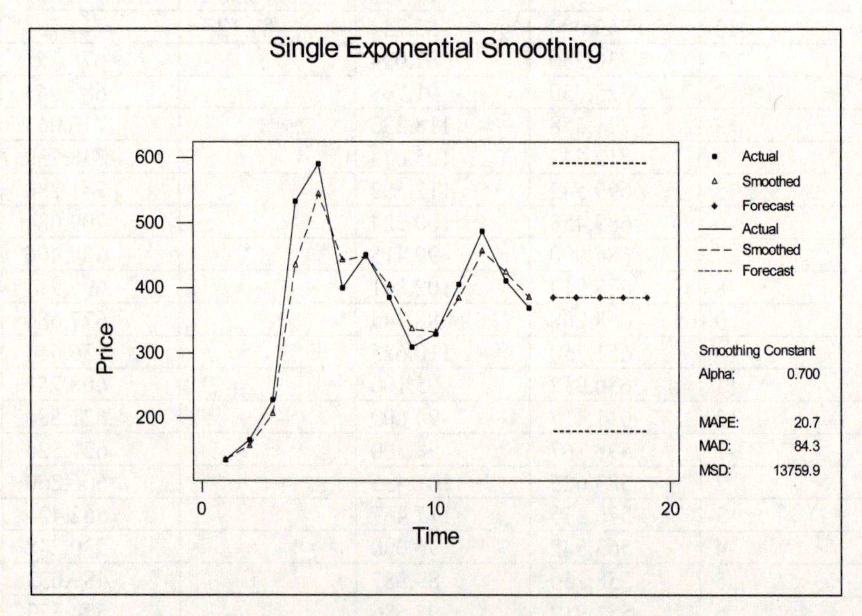

16.30 a. Obtain the smoothed series $\hat{x}_t$ as shown below.

$\hat{x}_1 = x_1$ and $\hat{x}_t = (1-\alpha)\hat{x}_{t-1} + \alpha x_t$

YEAR	Smoothed Value ($\alpha = 0.8$)	Smoothed Value ($\alpha = 0.6$)	Smoothed Value ($\alpha = 0.4$)	Smoothed Value ($\alpha = 0.2$)
1	50.20	50.20	50.20	50.20
2	37.16	40.42	43.68	46.94
3	23.91	28.53	34.45	41.67
4	25.10	26.65	30.83	38.42
5	31.34	30.40	31.66	37.31
6	31.31	30.94	31.51	36.11
7	21.30	23.66	26.43	32.65
8	15.86	18.16	21.66	29.02
9	17.17	17.76	19.99	26.72
10	22.31	21.27	21.44	26.09
11	20.94	20.87	21.10	24.99
12	21.47	21.31	21.30	24.32
13	26.37	25.08	23.82	24.97
14	34.47	31.93	28.89	27.28
15	46.33	42.35	37.06	31.68
16	45.59	44.18	40.39	34.43
17	37.60	39.03	38.48	34.66
18	31.60	33.67	35.13	33.75
19	26.72	28.77	31.28	32.10
20	23.26	24.95	27.73	30.16
21	14.49	17.36	21.56	26.59
22	19.62	19.48	21.29	25.45
23	38.16	33.47	29.90	28.92
24	76.67	65.17	52.46	40.40
25	74.21	70.23	60.91	47.04
26	53.00	56.71	55.63	47.17
27	55.88	56.64	56.02	49.06
28	53.66	54.52	54.85	49.86

b. The squared error, $e_t^2 = (x_t - \hat{x}_{t-1})^2$, for each forecast for each value of α is shown below.

YEAR	e_t^2 $(\alpha = 0.8)$	e_t^2 $(\alpha = 0.6)$	e_t^2 $(\alpha = 0.4)$	e_t^2 $(\alpha = 0.2)$
1				
2	265.69	265.69	265.69	265.69
3	274.23	392.83	532.69	693.80
4	2.21	9.78	81.87	264.78
5	60.80	39.05	4.29	30.44
6	0.00	0.81	0.13	36.17
7	156.45	147.38	161.66	299.68
8	46.26	83.83	142.29	329.39
9	2.69	0.44	17.28	132.69
10	41.32	34.05	13.00	9.71
11	2.94	0.44	0.70	30.17
12	0.43	0.54	0.25	11.52
13	37.59	39.61	39.68	10.79
14	102.54	130.36	160.76	132.89
15	219.79	301.61	416.47	484.98
16	0.87	9.28	69.63	188.18
17	99.74	73.64	22.98	1.38
18	56.21	79.79	70.16	20.80
19	37.20	66.80	92.65	68.04
20	18.66	40.57	78.77	94.07
21	120.21	159.96	237.94	318.94
22	41.05	12.54	0.43	32.34
23	537.38	543.65	462.55	301.03
24	2317.10	2790.64	3181.43	3292.49
25	9.44	71.08	447.01	1102.52
26	703.02	507.50	174.62	0.44
27	12.94	0.01	0.94	88.94
28	7.73	12.56	8.51	16.36
Total	5174.49	5814.44	6684.38	8258.23

Use the forecast with smoothing constant $\alpha = 0.8$ since it minimizes the sum of squared forecast errors.

16.32 If alpha is 1.0, then the forecast will always be equal to the first observation.

$\hat{X}_{t+h} = X_1$

16.34 **Double Exponential Smoothing**

```
Data          INDEX
Length        15.0000
NMissing      0

Smoothing Constants
Alpha (level): 0.7
Gamma (trend): 0.5

Accuracy Measures
MAPE:    65.69
MAD:     16.64
MSD:   1063.04
```

Row	Period	Forecast	Lower	Upper
1	16	127.69	86.93	168.46
2	17	142.91	91.44	194.38
3	18	158.12	95.02	221.22
4	19	173.33	98.10	248.57
5	20	188.54	100.89	276.20

Using Minitab, the forecasts for the next 5 years are 127.69, 142.91, 158.12, 173.33, and 188.54.

16.36 **Double Exponential Smoothing**

```
Data          FoodPrice
Length        14.0000
NMissing      0
Smoothing Constants
Alpha (level): 0.5
Gamma (trend): 0.5
Accuracy Measures
MAPE: 0.250818
MAD:  0.303748
MSD:  0.129733
```

Row	Time	FoodPrice	Smooth	Predict	Error
1	1	116.6	116.659	116.717	-0.117143
2	2	117.1	117.174	117.248	-0.147527
3	3	117.8	117.763	117.726	0.074162
4	4	118.9	118.617	118.334	0.566466
5	5	119.5	119.414	119.329	0.171002
6	6	120.3	120.235	120.169	0.130519
7	7	120.6	120.811	121.022	-0.422352
8	8	120.8	121.147	121.493	-0.693200
9	9	121.2	121.428	121.655	-0.455323
10	10	122.1	121.961	121.823	0.277445
11	11	122.6	122.513	122.426	0.174469
12	12	123.6	123.310	123.021	0.579363
13	13	124.2	124.082	123.963	0.236969
14	14	125.0	124.897	124.793	0.206530

Row	Period	Forecast	Lower	Upper
1	15	125.660	124.916	126.405
2	16	126.424	125.580	127.268
3	17	127.187	126.234	128.141

16.38 **Winters' multiplicative model**

```
Data          Sales
Length        24.0000
NMissing      0

Smoothing Constants
Alpha (level):      0.4
Gamma (trend):      0.5
Delta (seasonal): 0.6

Accuracy Measures
MAPE:     11.83
MAD:      29.65
MSD:   1545.12
```

Row	Period	Forecast	Lower	Upper
1	25	247.560	174.913	320.208
2	26	207.640	120.963	294.317
3	27	169.992	67.920	272.064
4	28	197.427	79.127	315.728
5	29	185.332	50.269	320.395
6	30	151.946	-0.236	304.129
7	31	121.120	-48.432	290.671
8	32	136.272	-50.828	323.371

Using Minitab, the forecasts for the next 8 quarters are 247.56, 207.64, 169.99, 197.43, 185.33, 151.95, 121.12, and 136.27. The graph of the data and the forecasts is shown below.

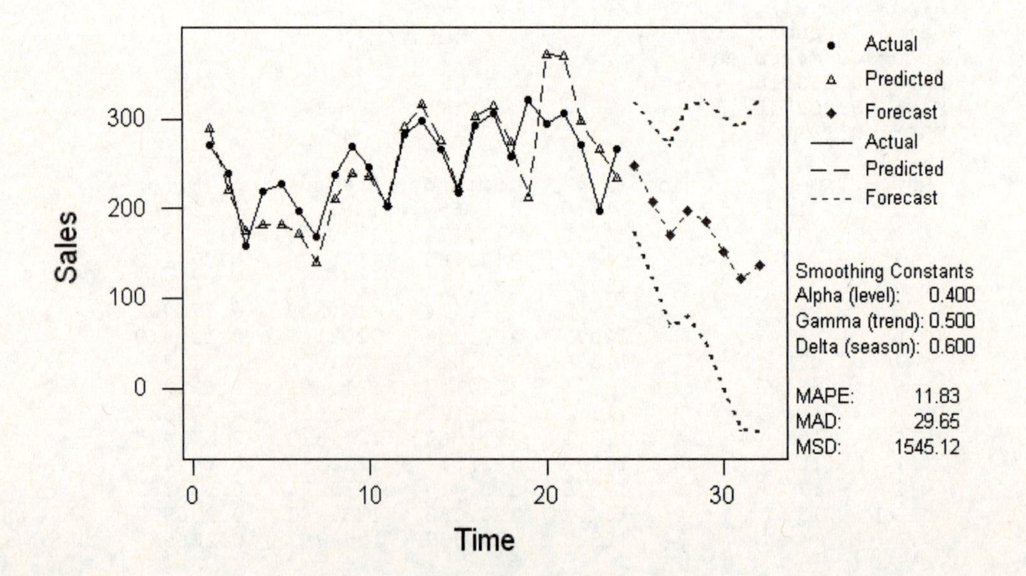

16.40 The data to be used for this exercise comes from Table 16.10: Index of Volume of Shares Traded:

Regression Analysis: Volume_Table16-10 versus Vol_lag1

```
The regression equation is
Volume_Table16-10 = 87.8 + 0.169 Vol_lag1
15 cases used 1 cases contain missing values

Predictor         Coef      SE Coef          T        P
Constant         87.85        30.45       2.88    0.013
Vol_lag1        0.1690       0.2855       0.59    0.564

S = 11.06       R-Sq = 2.6%      R-Sq(adj) = 0.0%
Analysis of Variance
Source            DF           SS          MS         F        P
Regression         1         42.9        42.9      0.35    0.564
Residual Error    13       1589.5       122.3
Total             14       1632.4
```

The first-order autoregressive model is

$\hat{y}_t = 87.85 + .169 y_{t-1} + a_t$

$y_{17} = 87.85 + .169(92) = 103.398$

$y_{18} = 87.85 + .169(103.398) = 105.324$

$y_{19} = 87.85 + .169(105.324) = 105.650$

$y_{20} = 87.85 + .169(105.650) = 105.705$

16.42 4th order model:

Regression Analysis: Starts versus Startlag1, Startlag2, ...

```
The regression equation is
Starts = 4.45 + 1.25 Startlag1 - 1.10 Startlag2 + 0.313 Startlag3
         - 0.064 Startlag4
20 cases used 4 cases contain missing values
Predictor         Coef      SE Coef          T        P
Constant         4.449        2.884       1.54    0.144
Startlag1       1.2517       0.2783       4.50     0.000
Startlag2      -1.0950       0.4182      -2.62     0.019
Startlag3       0.3131       0.4193       0.75     0.467
Startlag4      -0.0641       0.2935      -0.22     0.830

S = 1.042       R-Sq = 73.2%      R-Sq(adj) = 66.1%
```

$z - statistic\ for\ \phi_4 = -.218$. Fail to reject H_0 at the 10% level

3rd order model:

Regression Analysis: Starts versus Startlag1, Startlag2, ...

```
The regression equation is
Starts = 4.10 + 1.24 Startlag1 - 1.03 Startlag2 + 0.245 Startlag3
21 cases used 3 cases contain missing values
Predictor         Coef      SE Coef          T        P
Constant         4.100        2.183       1.88    0.078
Startlag1       1.2375       0.2509       4.93     0.000
Startlag2      -1.0325       0.2900      -3.56     0.002
Startlag3       0.2450       0.2695       0.91     0.376

S = 0.9808      R-Sq = 73.2%      R-Sq(adj) = 68.4%
```

$z - statistic\ for\ \phi_3 = .909$. Fail to reject H_0 at the 10% level

2nd order model:

Regression Analysis: Starts versus Startlag1, Startlag2

```
The regression equation is
Starts = 5.73 + 1.03 Startlag1 - 0.788 Startlag2
22 cases used 2 cases contain missing values
Predictor        Coef      SE Coef          T        P
Constant        5.728        1.245       4.60    0.000
Startlag       1.0332       0.1573       6.57    0.000
Startlag      -0.7877       0.1705      -4.62    0.000

S = 0.9765      R-Sq = 70.3%     R-Sq(adj) = 67.2%
```

$z-statistic\ for\ \phi_2 = -4.621$. Reject H_0 at the 10% level

1st order model:

Regression Analysis: Starts versus Startlag1

```
The regression equation is
Starts = 2.61 + 0.633 Startlag1
23 cases used 1 cases contain missing values

Predictor        Coef      SE Coef          T        P
Constant        2.614        1.451       1.80    0.086
Startlag       0.6333       0.1874       3.38    0.003

S = 1.376       R-Sq = 35.2%     R-Sq(adj) = 32.1%
```

Forecasts from the second order model:

$\hat{y}_{25} = 6.776, \hat{y}_{26} = 9.103, \hat{y}_{27} = 9.792, \hat{y}_{28} = 8.670, \hat{y}_{29} = 6.968$

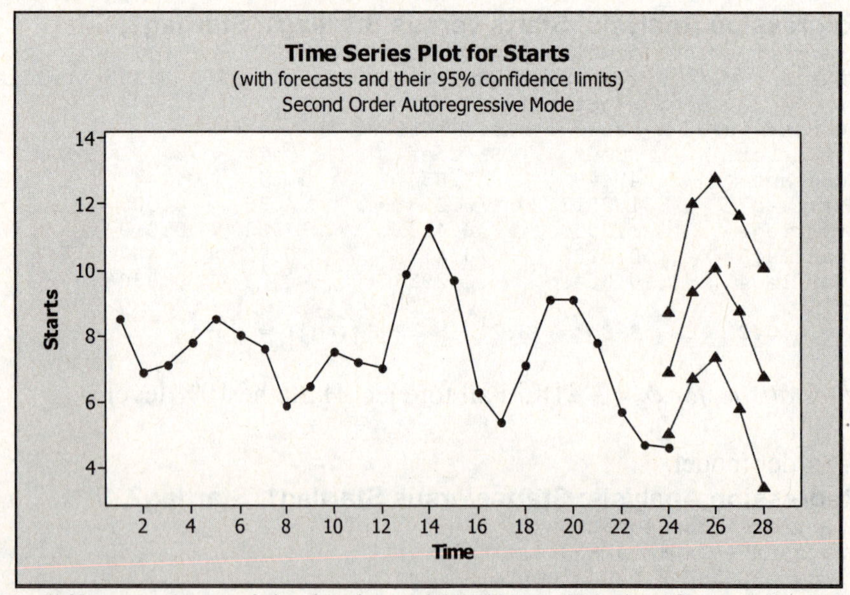

No difference in model selection with 10% or 5% significance level.

16.44 3rd order model:

Regression Analysis: Earnings versus esp_lag1, esp_lag2, ...

```
The regression equation is
Earnings = 2.48 + 0.870 esp_lag1 - 0.272 esp_lag2 - 0.093 esp_lag3
15 cases used 3 cases contain missing values
Predictor        Coef     SE Coef        T       P
Constant        2.477       1.317     1.88   0.087
esp_la         0.8697      0.3030     2.87   0.015
esp_la        -0.2720      0.3928    -0.69   0.503
esp_la        -0.0932      0.3076    -0.30   0.768
S = 1.328     R-Sq = 51.7%     R-Sq(adj) = 38.5%
```

$z-statistic\ for\ \phi_3 = -.303$, Fail to reject H_0 at the 10% level

2nd order model:

Regression Analysis: Earnings versus esp_lag1, esp_lag2

```
The regression equation is
Earnings = 2.16 + 0.916 esp_lag1 - 0.349 esp_lag2
16 cases used 2 cases contain missing values
Predictor        Coef     SE Coef        T       P
Constant        2.158       1.033     2.09   0.057
esp_la         0.9161      0.2608     3.51   0.004
esp_la        -0.3489      0.2629    -1.33   0.207
S = 1.237     R-Sq = 52.4%     R-Sq(adj) = 45.0%
```

$z-statistic\ for\ \phi_2 = -1.327$, Fail to reject H_0 at the 10% level

1st order model:

Regression Analysis: Earnings versus esp_lag1

```
The regression equation is
Earnings = 1.57 + 0.696 esp_lag1
17 cases used 1 cases contain missing values
Predictor        Coef     SE Coef        T       P
Constant       1.5697      0.9342     1.68   0.114
esp_la         0.6962      0.1900     3.66   0.002
S = 1.234     R-Sq = 47.2%     R-Sq(adj) = 43.7%
```

$z-statistic\ for\ \phi_1 = 3.664$, Reject H_0 at the 10% level

Use the 1st order model for forecasting

$\hat{y}_{19} = 5.927,\ \hat{y}_{20} = 5.695,\ \hat{y}_{21} = 5.534,\ \hat{y}_{22} = 5.422$

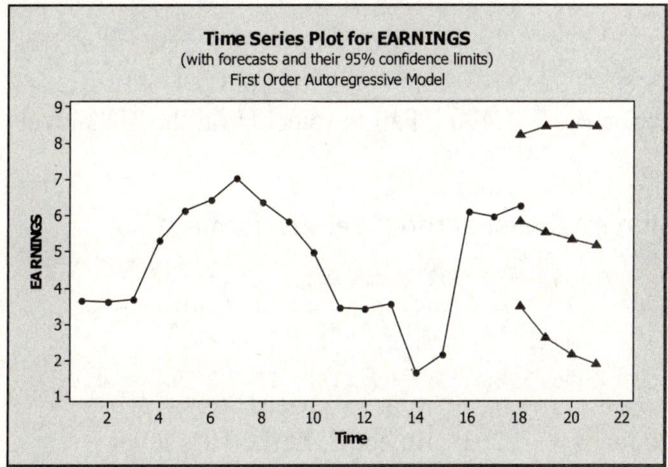

No difference in model selection with 10% or 5% significance level.

16.46 $\hat{X}_{1996} = 202 + 1.1(951) - .48(923) + .17(867) = 952.45$

$\hat{X}_{1997} = 202 + 1.1(952.45) - .48(951) + .17(923) = 950.13$

$\hat{X}_{1998} = 202 + 1.1(950.13) - .48(952.45) + .17(951) = 951.64$

16.48 4th order model:

Regression Analysis: earndiff versus earnlag1, earnlag2, ...

```
The regression equation is
earndiff = 0.0941 - 0.936 earnlag1 - 0.547 earnlag2
           - 0.367 earnlag3 - 0.251 earnlag4
19 cases used 5 cases contain missing values
Predictor        Coef      SE Coef        T       P
Constant      0.09409      0.03251     2.89   0.012
earnlag1      -0.9358       0.2647    -3.53   0.003
earnlag2      -0.5471       0.3451    -1.59   0.135
earnlag3      -0.3674       0.2905    -1.26   0.227
earnlag4      -0.2514       0.2122    -1.19   0.256
S = 0.06788     R-Sq = 51.4%     R-Sq(adj) = 37.6%
```

T-statistic for $\phi_4 = -1.185$. Fail to reject H_0 at the 10% level

3rd order model:

Regression Analysis: earndiff versus earnlag1, earnlag2, ...

```
The regression equation is
earndiff = 0.0799 - 0.907 earnlag1 - 0.479 earnlag2
           - 0.173 earnlag3
20 cases used 4 cases contain missing values
Predictor        Coef      SE Coef        T       P
Constant      0.07994      0.02358     3.39   0.004
earnlag1      -0.9068       0.2466    -3.68   0.002
earnlag2      -0.4787       0.2601    -1.84   0.084
earnlag3      -0.1735       0.2051    -0.85   0.410
S = 0.06707     R-Sq = 46.9%     R-Sq(adj) = 37.0%
```

T-statistic for $\phi_3 = -.846$. Fail to reject H_0 at the 10% level

2nd order model:

Regression Analysis: earndiff versus earnlag1, earnlag2

```
The regression equation is
earndiff = 0.0596 - 0.704 earnlag1 - 0.286 earnlag2
21 cases used 3 cases contain missing values
Predictor        Coef      SE Coef        T       P
Constant      0.05961      0.01742     3.42   0.003
earnlag1      -0.7037       0.1943    -3.62   0.002
earnlag2      -0.2865       0.1923    -1.49   0.154
S = 0.06676     R-Sq = 43.4%     R-Sq(adj) = 37.2%
```

T-statistic for $\phi_2 = -1.490$. Fail to reject H_0 at the 10% level

1st order model:

Regression Analysis: earndiff versus earnlag1

```
The regression equation is
earndiff = 0.0399 - 0.592 earnlag1
22 cases used 2 cases contain missing values
Predictor        Coef      SE Coef        T       P
Constant      0.03995      0.01757     2.27   0.034
earnlag1      -0.5920       0.1814    -3.26   0.004
S = 0.07858     R-Sq = 34.7%     R-Sq(adj) = 31.5%
```

T-statistic for $\phi_1 = -3.263$. Reject H_0 at the 10% level

Use the 1st order model for forecasting

$\hat{y}_{25} = .070,\ \hat{y}_{26} = -.001,\ \hat{y}_{27} = .041$

16.50 Parts a,b,c:

Period	a. Unweighted	b. Laspeyres Price	c. Laspeyres Quantity
1	100.00	100.00	100.00
2	110.30	109.72	103.78
3	117.43	115.55	100.04
4	127.52	123.47	105.29
5	143.96	137.18	106.00
6	158.22	149.41	105.85

16.52 Forecasts are generated by analyzing each individual component; trend, seasonal and cyclical. Then, once each component has been analyzed and measured, the information is then incorporated into the forecasting model.

16.54 A seasonally adjusted time series is one that is free from the effects of seasonal influence. Government agencies expend large efforts on seasonal adjustments in order to gain a clearer picture of the underlying data pattern.

16.56 a. **Runs Test: Sales**

```
Sales
K =    737.0000

    The observed number of runs =  10
    The expected number of runs =  13.0000
    12 Observations above K   12 below
              The test is significant at  0.2105
              Cannot reject at alpha = 0.05
```

b.

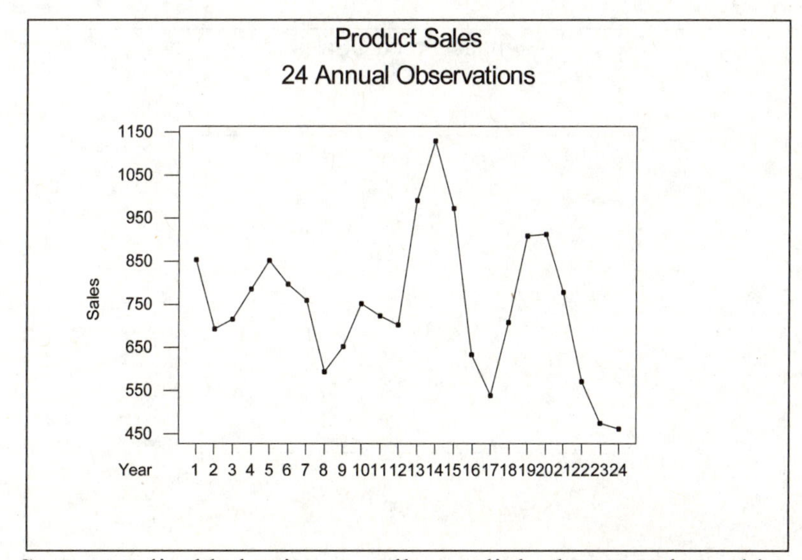

Strong cyclical behavior as well as a slight downward trend is exhibited.

c.

Year	Sales	5pt MA
1	853	*
2	693	*
3	715	779.4
4	785	768.2
5	851	781.2
6	797	756.8
7	758	729.8
8	593	709.8
9	650	695.0
10	751	683.8
11	723	763.4
12	702	859.2
13	991	903.4
14	1129	885.0
15	972	852.2
16	631	795.6
17	538	751.2
18	708	739.2
19	907	768.4
20	912	774.6
21	777	727.6
22	569	638.0
23	473	*
24	459	*

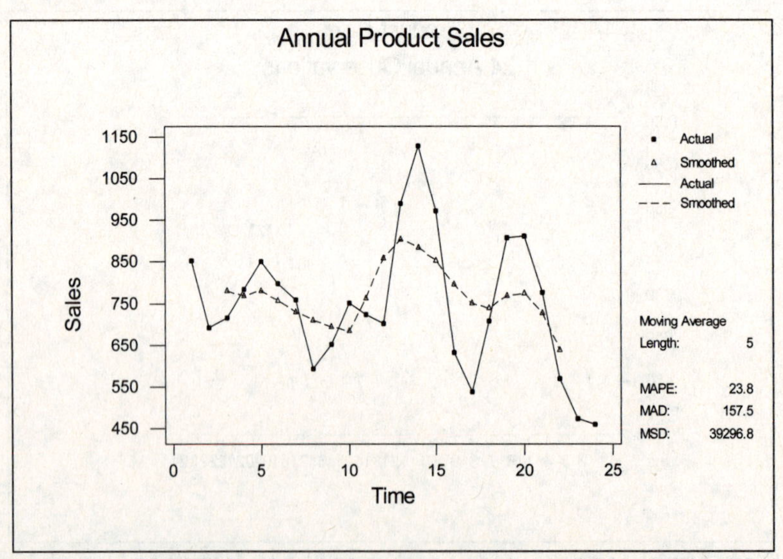

Strong cyclical behavior and a downward trend is exhibited.

16.58 a. **Moving average**

```
Data        PriceIndex
Length      15.0000
NMissing    0

Moving Average
Length: 3

Accuracy Measures
MAPE:   5.6999
MAD:    6.1389
MSD:   84.4352
```

Row	Period	PriceIndex	AVER3	Predict	Error
1	1	79	*	*	*
2	2	87	85.000	*	*
3	3	89	88.667	*	*
4	4	90	89.000	85.000	5.0000
5	5	88	89.000	88.667	-0.6667
6	6	89	90.333	89.000	0.0000
7	7	94	91.667	89.000	5.0000
8	8	92	91.333	90.333	1.6667
9	9	88	92.000	91.667	-3.6667
10	10	96	100.333	91.333	4.6667
11	11	117	109.667	92.000	25.0000
12	12	116	115.667	100.333	15.6667
13	13	114	114.333	109.667	4.3333
14	14	113	112.000	115.667	-2.6667
15	15	109	*	114.333	-5.3333

b.

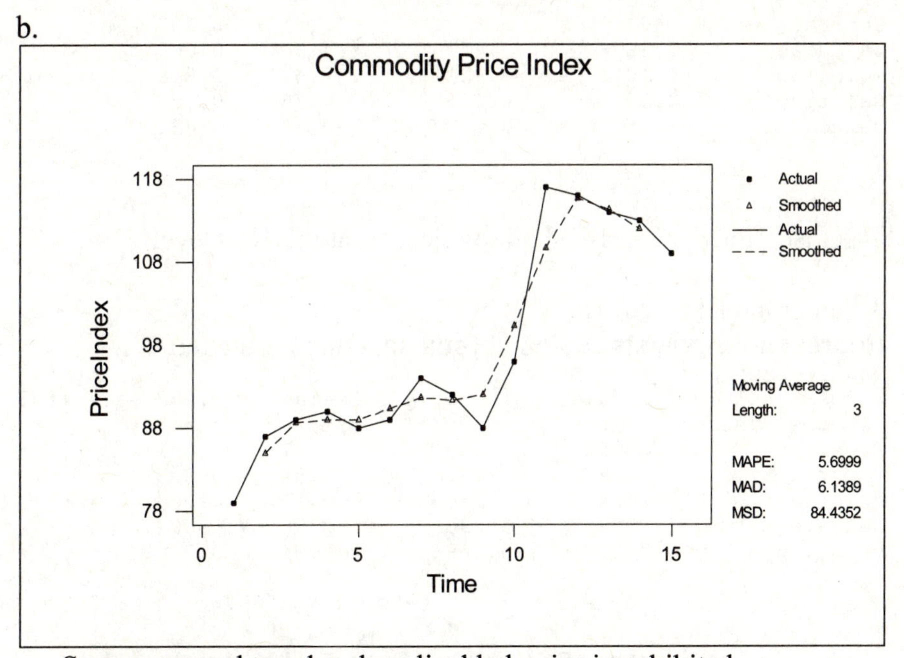

Strong upward trend and cyclical behavior is exhibited.

16.60 **Double Exponential Smoothing for PRICE**

```
Data     PRICE
Length  15

Smoothing Constants
Alpha (level)  0.7
Gamma (trend)  0.6

Accuracy Measures
MAPE    5.9448
MAD     6.1265
MSD    63.3177

Period   Forecast    Lower     Upper
16       108.2722  93.2624  123.2819
17       105.5983  86.6483  124.5484
18       102.9245  79.6921  126.1570
19       100.2507  72.5519  127.9496
```

Using Minitab, the forecasts for the next four months are 108.27, 105.60, 102.92, and 100.25.

16.62 4th order model

Regression Analysis: Sales versus saleslag1, saleslag2, ...

```
The regression equation is
Sales = 435 + 1.26 saleslag1 - 1.10 saleslag2 + 0.321 saleslag3
          - 0.063 saleslag4
20 cases used 4 cases contain missing values
Predictor        Coef     SE Coef          T        P
Constant        434.9       287.5       1.51    0.151
saleslag       1.2626      0.2796       4.52    0.000
saleslag      -1.1016      0.4211      -2.62    0.019
saleslag       0.3210      0.4220       0.76    0.459
saleslag      -0.0634      0.2938      -0.22    0.832

S = 103.9       R-Sq = 73.3%     R-Sq(adj) = 66.2%
```

T-statistic for $\phi_4 = -.216$. Fail to reject H_0 at the 10% level.

3rd order model

Regression Analysis: Sales versus saleslag1, saleslag2, ...

```
The regression equation is
Sales = 401 + 1.25 saleslag1 - 1.04 saleslag2 + 0.253 saleslag3
21 cases used 3 cases contain missing values
Predictor        Coef     SE Coef          T        P
Constant        400.7       217.8       1.84    0.083
saleslag       1.2484      0.2509       4.97    0.000
saleslag      -1.0394      0.2905      -3.58    0.002
saleslag       0.2532      0.2694       0.94    0.360

S = 97.76       R-Sq = 73.3%     R-Sq(adj) = 68.6%
```

T-statistic for $\phi_3 = .940$. Fail to reject H_0 at the 10% level.

2nd order model

Regression Analysis: Sales versus saleslag1, saleslag2

```
The regression equation is
Sales = 568 + 1.04 saleslag1 - 0.785 saleslag2
22 cases used 2 cases contain missing values
Predictor        Coef     SE Coef          T        P
Constant        568.1       124.8       4.55    0.000
saleslag       1.0368      0.1580       6.56    0.000
saleslag      -0.7847      0.1710      -4.59    0.000

S = 97.72       R-Sq = 70.2%      R-Sq(adj) = 67.1%
```

T-statistic for $\phi_2 = -4.590$. Reject H_0 at the 10% level.

Use the 2nd order model for forecasting.

$\hat{y}_{25} = 672.829$, $\hat{y}_{26} = 905.554$, $\hat{y}_{27} = 979.039$

Chapter 17:

Additional Topics in Sampling

17.2 a. $\bar{x}_3 = 43.3$, $\hat{\sigma}^2_{\bar{x}_3} = \frac{s^2}{n}\frac{N-n}{N} = \frac{(12.3)^2}{50}\frac{208-50}{208} = 2.2984$

90% confidence interval: $43.3 \pm 1.645\sqrt{2.2984}$: 40.806 up to 45.794

b. $\bar{x}_{st} = \frac{1}{N}\sum_{j=1}^{k} N_j\bar{x}_j = \frac{152(27.6)+127(39.2)+208(43.3)}{487} = 37.3306$

c. $\hat{\sigma}^2_{\bar{x}_1} = \frac{(7.1)^2}{40}\frac{152-40}{152} = .9286$, $\hat{\sigma}^2_{\bar{x}_2} = \frac{(9.9)^2}{40}\frac{127-40}{127} = 1.6785$

$$\hat{\sigma}^2_{\bar{x}_{st}} = \frac{(152)^2(.9286)+(127)^2(1.6785)+(208)^2(2.2984)}{(487)^2} = .6239$$

90% confidence interval: $37.3306 \pm 1.645\sqrt{.6239}$: 36.0313 up to 38.6299

95% confidence interval: $37.3306 \pm 1.96\sqrt{.6239}$: 35.7825 up to 38.8787

17.4 a. $\hat{\sigma}^2_{\bar{x}_1} = \frac{s^2}{n}\frac{N-n}{N} = \frac{(1.04)^2}{50}\frac{632-50}{632} = .0199$; $3.12 \pm 1.96\sqrt{.0199}$:
2.8435 up to 3.3965

b. $\hat{\sigma}^2_{\bar{x}_2} = \frac{(.86)^2}{50}\frac{529-50}{529} = .0134$; $3.37 \pm 1.96\sqrt{.0134}$: 3.1431 up to 3.5969

c. $\hat{\sigma}^2_{\bar{x}_{st}} = \frac{(632)^2(.0199)+(529)^2(.0134)}{(1161)^2} = .0087$; $\bar{x}_{st} = 3.2339$

$3.2339 \pm 1.96\sqrt{.0087}$: 3.0513 up to 3.4166

17.6 a. $N\bar{x}_{st} = 237(120) + 198(150) + 131(180) = 81{,}720$

b. $\hat{\sigma}^2_{\bar{x}_1} = \frac{s^2}{n}\frac{N-n}{N} = \frac{93^2}{40}\frac{120-40}{120} = 144.15$, $\hat{\sigma}^2_{\bar{x}_2} = \frac{64^2}{45}\frac{150-45}{150} = 63.7156$,

$\hat{\sigma}^2_{\bar{x}_3} = \frac{47^2}{50}\frac{180-50}{180} = 31.9078$

$$\hat{\sigma}^2_{\bar{x}_{st}} = \frac{(120)^2(144.15)+(150)^2(63.71556)+(180)^2(31.9078)}{(450)^2} = 22.4354$$

95% confidence interval: $181.6(450) \pm 1.96\sqrt{22.4354}\,(450)$:
$77{,}542.3153 < N\mu < 85{,}897.6847$

17.8 a. $\hat{p}_{st} = [100\frac{6}{25} + 50\frac{14}{25}]/150 = .3467$

b. $\hat{\sigma}^2_{\hat{p}_1} = \frac{\hat{p}_1(1-\hat{p}_1)}{n_1 - 1}\frac{N_1 - n_1}{N_1} = \frac{.24(.76)}{25-1}\frac{100-25}{100} = .0057$

$\hat{\sigma}^2_{\hat{p}_2} = \frac{.56(.44)}{25-1}\frac{50-25}{50} = .0051,$

$\hat{\sigma}^2_{\hat{p}_{st}} = \frac{(100)^2(.0057) + (50)^2(.0051)}{(150)^2} = .0031$

90% confidence interval: $.3467 \pm 1.645\sqrt{.0031}$: .2550 up to .4383

95% confidence interval: $.3467 \pm 1.96\sqrt{.0031}$: .2375 up to .4559

17.10 a. $n_3 = \frac{208}{487}130 = 55.52 = 56$ observations

b. $n_3 = \left[\frac{208(12.3)}{152(7.1) + 127(9.9) + 208(12.3)}\right]130 = 67.95 = 68$ observations

17.12 a. $n_1 = \frac{632}{1161}100 = 54.43 = 55$ observations

b. $n_1 = \left[\frac{632(1.04)}{632(1.04) + 529(.86)}\right]100 = 59.09 = 60$ observations

17.14 a. $n_2 = \frac{1031}{1395}100 = 73.91 = 74$ observations

b. $n_2 = \left[\frac{1031(219.9)}{364(87.3) + 1031(219.9)}\right]100 = 87.71 = 88$ observations

17.16 Proportional allocation: $\sigma_{\bar{x}} = \frac{500}{1.96} = 255.1020$

$\sum N_j \sigma^2{}_j = 1150(4000)^2 + 2120(6000)^2 + 930(800000)^2 = 15424 \times 10^7$

$n = \frac{15424 \times 10^7}{4200(255.1020)^2 + 15424 \times 10^7 / 4200} = 497.47 = 498$ observations

Optimal allocation:

$\sum N_j \sigma_j = 1150(4000) + 2120(6000) + 930(8000) = 24760000$

$n = \frac{(24760000)^2 / 4200}{4200(255.1020)^2 + 15424 \times 10^7 / 4200} = 470.78 = 471$ observations

17.18 a. $\bar{x}_c = \dfrac{69(83)+75(64)+\cdots+71(98)}{497} = 91.6761$

b. $\hat{\sigma}^2_{\bar{x}_c} = \dfrac{(52-8)}{52(8)(61.125)^2}\dfrac{(69)^2(83-91.67605634)^2+\cdots+(71)^2(98-91.67605634)^2}{8-1}$

$= 66.409$

99% confidence interval: $91.6761 \pm 2.58\sqrt{66.4090}$

70.6920 up to 112.6602

17.20 a. $\hat{p}_c = \dfrac{24+\cdots+34}{497} = .4507$

b. $\hat{\sigma}^2_{\hat{p}_c} = \dfrac{52-8}{52(8)(62.125)^2}\dfrac{(69)^2\left(\dfrac{24}{69}-.4507\right)^2+\cdots+(71)^2\left(\dfrac{34}{71}-.4507\right)^2}{8-1} = .0013$

95% confidence interval: $.4507 \pm 1.96\sqrt{.0013}$: .38 up to .5214

17.22 $\sigma_{\bar{x}} = \dfrac{5000}{1.645} = 3039.5$, $n = \dfrac{720(37600)^2}{719(3039)^2+(37600)^2} = 126.34 = 127$ observations.

Additional sample observations needed are 127-20 = 107

17.24 $\sigma_{\bar{x}} = \dfrac{20}{1.96} = 10.2$

$$n = \frac{(100(105)+180(162)+200(183))^2/480}{480(10.2)^2+(100(105)^2+180(162)^2+200(183)^2)/480} = 159.35 = 160$$

observations. Additional sample observations needed: 160-30 = 130

17.26 a. $\hat{p} = \dfrac{38}{61} = .623$, $\hat{\sigma}^2_{\hat{p}} = \dfrac{.623(.377)}{61}\dfrac{100-61}{100} = .0015$

90% confidence interval: $.623 \pm 1.645\sqrt{.0015}$: .559 up to .687

b. If the sample information is not randomly selected, the resulting conclusions may be biased

17.28 a. $\bar{x}_{st} = \dfrac{120(1.6)+180(.74)}{300} = \dfrac{325.2}{300} = 1.084$

$\hat{\sigma}^2_{\bar{x}_1} = \dfrac{s^2}{n}\dfrac{N-n}{N} = \dfrac{(.98)^2}{20}\dfrac{120-20}{120} = .0400$

$\hat{\sigma}^2_{\bar{x}_2} = \dfrac{(.56)^2}{20}\dfrac{180-20}{180} = .0139$, $\hat{\sigma}^2_{\bar{x}_{st}} = \dfrac{(120)^2(.04)+(180)^2(.0139)}{(300)^2} = .0114$

95% confidence interval for mean number of errors per page in the book:

$1.084 \pm 1.96\sqrt{.0114}$: .8747 up to 1.2933

b. $N\bar{x}_{st} - Z_{\alpha/2} N\hat{\sigma}_{\bar{x}_{st}} < N\mu < N\bar{x}_{st} + Z_{\alpha/2} N\hat{\sigma}_{\bar{x}_{st}}$

where, $N\bar{x}_{st} = (120)(1.6) + (180)(.74) = 325.2$

$N^2\hat{\sigma}^2_{\bar{x}_{st}} = (120)^2.0400 + (180)^2.0139 = 1{,}029.3408$

$N\hat{\sigma}_{\bar{x}_{st}} = \sqrt{1029.3408} = 32.08334$, $325.2 \pm 2.58(32.08334)$

99% confidence interval for the total number of errors in the book: 242.4279 up to 407.9721 or from 243 total errors up to 408 total errors.

17.30 a. $n_1 = \frac{352}{970} 80 = 29.03 = 30$ observations

b. $n_1 = \left[\frac{352(4.9)}{352(4.9) + 287(6.4) + 331(7.6)} \right] 80 = 22.7 = 23$ observations

17.32 Refer to section 17.1 – Stratified Sampling

Chapter 18:

Statistical Decision Theory

18.2 D is dominated by C. Therefore, D is inadmissible.

18.4 a. Note – D is dominated by C. Hence D is inadmissible
Maximin criterion would select production process C:

Actions	States of Nature			
Prod. Process	Low Demand	Moderate Demand	High Demand	Min Payoff
A	100,000	350,000	900,000	100,000
B	150,000	400,000	700,000	150,000
C	250,000	400,000	600,000	250,000

b. Minimax regret criterion would select production process A:

Actions	States of Nature			
Prod. Process	Low Demand	Moderate Demand	High Demand	Max Regret
A	150,000	50,000	0	150,000
B	100,000	0	200,000	200,000
C	0	0	300,000	300,000

18.6

Actions	States of Nature			
Prod. Process	Low Demand	Moderate Demand	High Demand	Min Payoff
A	70,000	120,000	200,000	70,000
B	80,000	120,000	180,000	80,000
C	100,000	125,000	160,000	100,000
D*	100,000	120,000	150,000	Inadmissible
E	60,000	115,000	220,000	60,000

*inadmissible

Therefore, production process C would be chosen using the Maximin Criterion

Actions Regrets or Opportunity Loss Table

Prod. Process	Low Demand	Moderate Demand	High Demand	Max Regret
A	30,000	5,000	20,000	30,000
B	20,000	5,000	40,000	40,000
C	0	0	60,000	60,000
D*				Inadmissible
E	40,000	10,000	0	40,000

*inadmissible

Therefore, production process A would be chosen using the Minimax Regret Criterion

18.8 Assume a situation with two states of nature and two actions. Let both actions be admissible. The payoff Matrix is:

Action	S1	S2
A1	M_{11}	M_{12}
A2	M_{21}	M_{22}

Then action A1 will be chosen by both the Maximin and the Minimax Regret Criteria if for: $M_{11} > M_{21}$ and $M_{12} < M_{22}$ and $(M_{11} - M_{21}) > (M_{22} - M_{12})$

18.10 a. Payoff table for students' decision-making problem

Actions	Offered Better Position	Not Offered Better Position
Interview	4500	-500
Don't Interview	0	0

b. EMV(Interview) = .05(4500) + .95(-500) = -250
EMV(Don't Interview) = 0
Therefore, the optimal action: Don't Interview

18.12 a. EMV(Certificate of Deposit) = 1200
EMV(Low risk stock fund) = .2(4300) + .5(1200) + .3(-600) = 1280
EMV(High risk stock fund) = .2(6600) + .5(800) + .3(-1500) = 1270
Therefore, the optimal action: Low risk stock fund

b. Decision tree

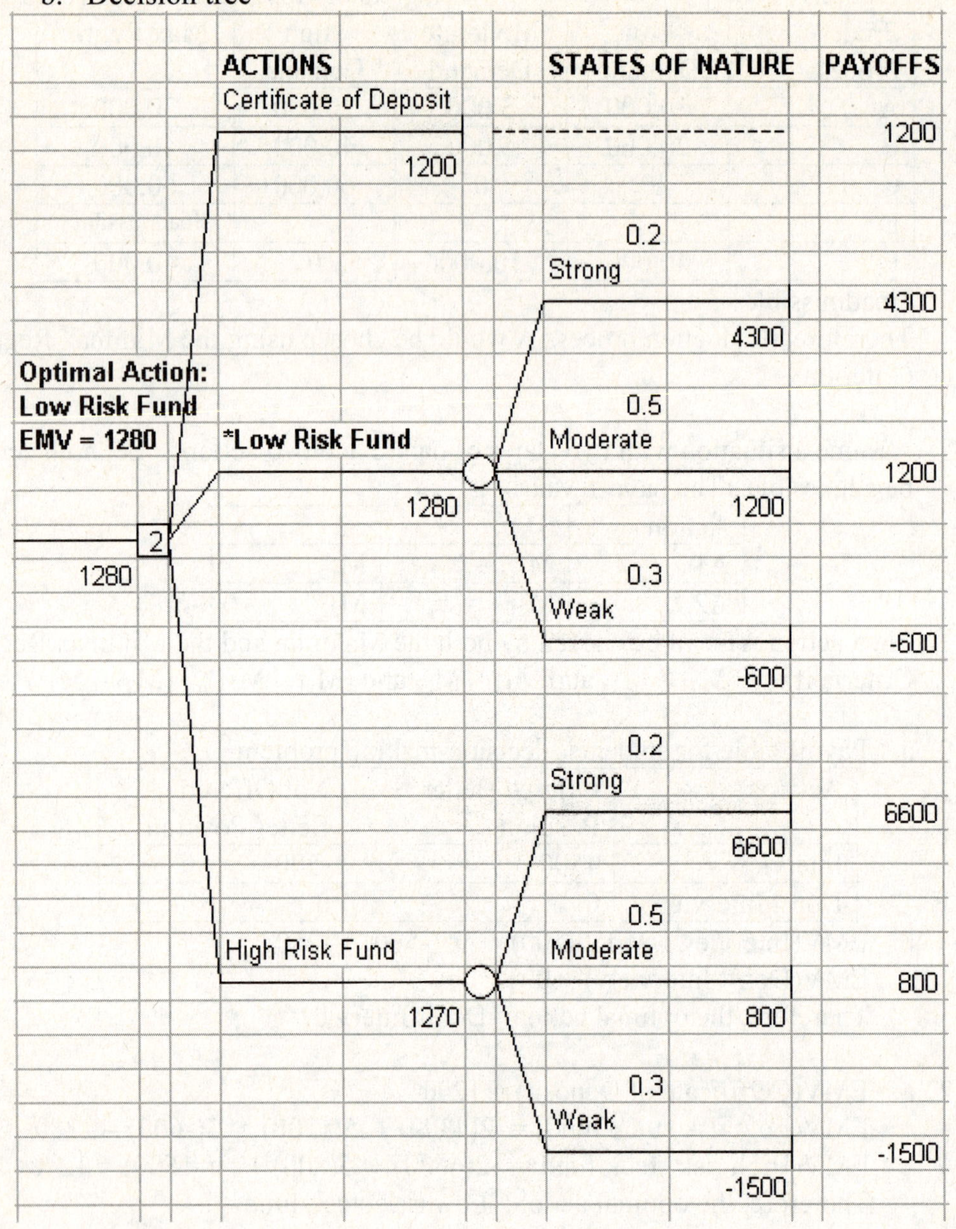

18.14 a. i) false; ii) true; iii) true

b. No

18.16 a. EMV(New) = .4(130,000) + .4(60,000) + .2(-10,000) = 74,000

EMV(Old) = .4(30,000) + .4(70,000) + .2(90,000) = 58,000

Therefore, the optimal action: New center

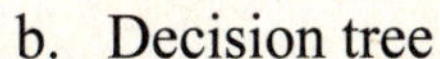

b. Decision tree

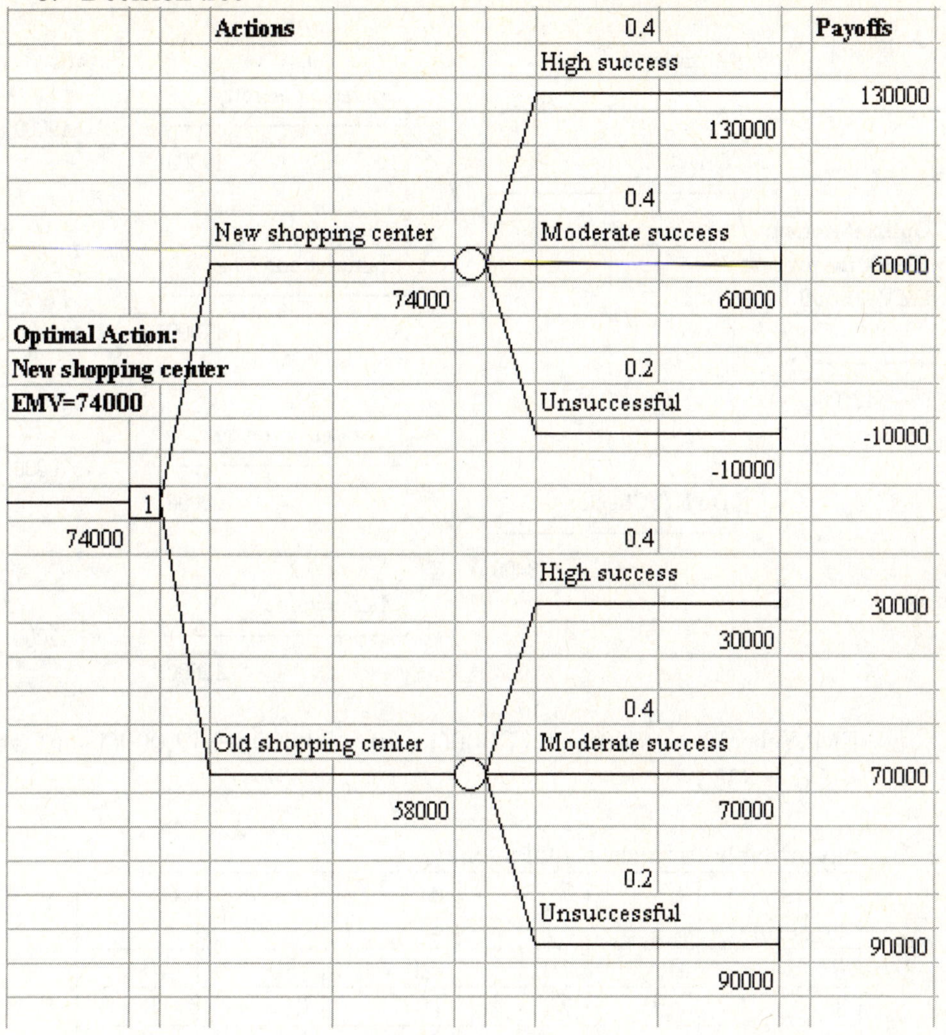

18.18 a. $EMV(A) = 30{,}000 + 350{,}000p + 900{,}000(.7 - p) = 660{,}000 - 550{,}000p$
$EMV(B) = 45{,}000 + 400{,}000p + 700{,}000(.7 - p) = 535{,}000 - 300{,}000p$
$EMV(C) = 75{,}000 + 400{,}000p + 600{,}000(.7 - p) = 495{,}000 - 200{,}000p$
$EMV(D) = 75{,}000 + 400{,}000p + 550{,}000(.7 - p) = 460{,}000 - 150{,}000p$
$EMV(A) = 660 - 550p > 535 - 300p = EMV(B)$ when $p < .5$
$EMV(A) = 660 - 550p > 495 - 200p = EMV(C)$ when $p < .471$
$EMV(A) = 660 - 550p > 460 - 150p = EMVA(D)$ when $p < .5$
For $p < .471$, the EMV criterion chooses action A, same decision as in 18.13. Note that D was "inadmissible"

b. $EMV(A) = 170{,}000 + .3a > 415{,}000 = EMV(B) > EMV(C) > EMV(D)$ when $a > 816{,}667$

18.20 a. $EMV(\text{check}) = .8(20{,}000 - 1{,}000) + .2(20{,}000 - 1{,}000 - 2{,}000) = 18{,}600$
$EMV(\text{not check}) = .8(20{,}000) + .2(12{,}000) = 18{,}400$
Therefore, the optimal action: Check the process

b. Decision tree

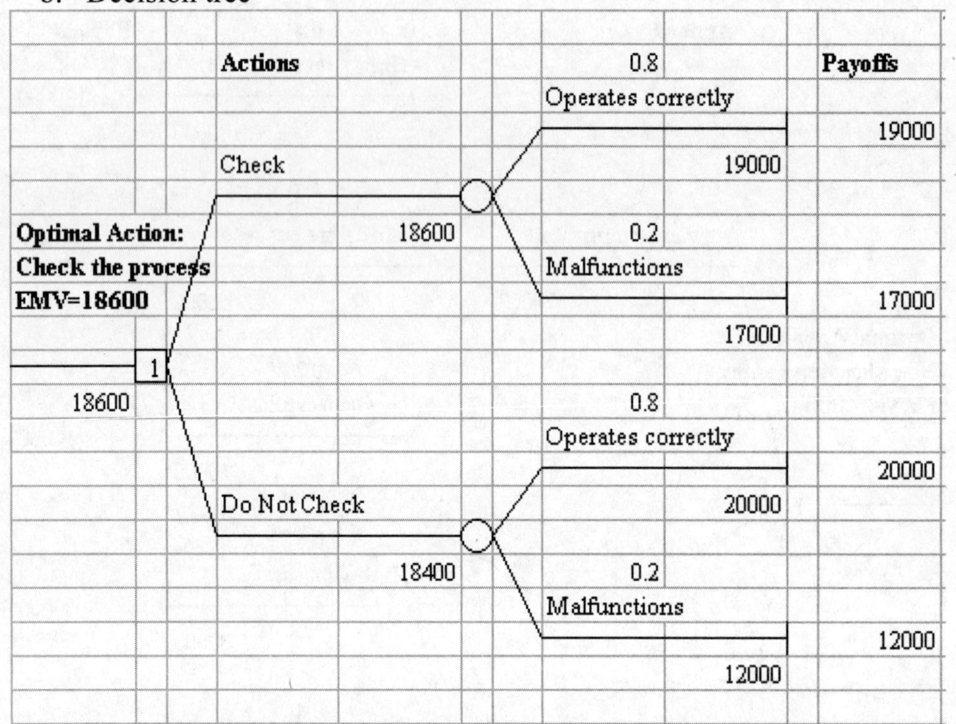

c. EMV(check) = 19,000p + 17,000(1 – p) > 20,000p + 12,000(1 – p) when p < 5/6 = .8333

18.22 a. payoff table for a car rental agency

Extra Ordering	6	7	8	9	10
0	0	-10	-20	-30	-40
1	-20	20	10	0	-10
2	-40	0	40	30	20
3	-60	-20	20	60	50
4	-80	-40	0	40	80

b. Per the EMV criterion, the optimal action is to order 2 extra cars:
Extra Orders

	6	7	8	9	10	EMV
0	0(.1)	-10(.3)	-20(.3)	-30(.2)	-40(.1)	-19
1	-20(.1)	20(.3)	10(.3)	0(.2)	-10(.1)	6
2	-40(.1)	0(.3)	40(.3)	30(.2)	20(.1)	16
3	-60(.1)	-20(.3)	20(.3)	60(.2)	50(.1)	11
4	-80(.1)	-40(.3)	0(.3)	40(.2)	80(.1)	-4

18.24 a. Action A1 is taken if $M_{11}p + M_{12}(1-p) > M_{21}p + (1-p)M_{22}$ or $p(M_{11} - M_{21}) > (1-p)(M_{22} - M_{12})$

b. Action A1 inadmissible implies that A1 will be chosen only if $p > 1$. In short, for part a. to be true, both payoffs of A1 cannot be less than the corresponding payoffs of A2.

18.26 a. Optimal action per the EMV criterion is action A (see answer to 18.13)

b. $P(L \mid P) = \dfrac{(.5)(.3)}{(.5)(.3)+(.3)(.4)+(.1)(.3)} = \dfrac{.15}{.3} = .5$

$P(M \mid P) = \dfrac{(.4)(.3)}{(.5)(.3)+(.3)(.4)+(.1)(.3)} = \dfrac{.12}{.3} = .4$

$P(H \mid P) = \dfrac{(.1)(.3)}{(.5)(.3)+(.3)(.4)+(.1)(.3)} = \dfrac{.03}{.3} = .1$

c. EMV(A) = .5(100,000)+.4(350,000)+.1(900,000) = 280,000
EMV(B) = .5(150,000)+.4(400,000)+.1(700,000) = 305,000
EMV(C) = .5(250,000)+.4(400,000)+.1(600,000) = 345,000
Therefore, the optimal action: C

d. $P(L \mid F) = \dfrac{(.3)(.3)}{(.3)(.3)+(.4)(.4)+(.2)(.3)} = \dfrac{.09}{.31} = .2903$

$P(M \mid F) = \dfrac{(.4)(.4)}{(.3)(.3)+(.4)(.4)+(.2)(.3)} = \dfrac{.16}{.31} = .5161$

$P(H \mid F) = \dfrac{(.2)(.3)}{(.3)(.3)+(.4)(.4)+(.2)(.3)} = \dfrac{.06}{.31} = .1935$

e. EMV(A) = .2903(100,000)+.5161(350,000)+.1935 (900,000) = 383,815
EMV(B) = .2903(150,000)+.5161(400,000)+.1935(700,000) = 385,435
EMV(C) = .2903(250,000)+.5161(400,000)+.1935(600,000) = 395,115
Therefore, the optimal action: C

f. $P(L \mid G) = \dfrac{(.2)(.3)}{(.2)(.3)+(.3)(.4)+(.7)(.3)} = \dfrac{.06}{.39} = .1538$

$P(M \mid G) = \dfrac{(.3)(.4)}{(.2)(.3)+(.3)(.4)+(.7)(.3)} = \dfrac{.12}{.39} = .3077$

$P(H \mid G) = \dfrac{(.7)(.3)}{(.2)(.3)+(.3)(.4)+(.7)(.3)} = \dfrac{.21}{.39} = .5385$

g. EMV(A) = .1538(100,000)+.3077(350,000)+.5385(900,000) = 607,692
EMV(B) = .1538(150,000)+.3077(400,000)+.5385(700,000) = 523,077
EMV(C) = .1538(250,000)+.3077(400,000)+.5385(600,000) = 484,615
Therefore, the optimal action: A

18.28 a. $P(E \mid P) = \frac{(.8)(.6)}{(.8)(.6)+(.1)(.4)} = \frac{.48}{.52} = .9231$

P(not E|P) = 1 – P(E|P) = 1 - .9231 = .0769

b. EMV(S) = .9231(50,000)+.0769(50,000) = 50,000

EMV(R) = .9231(125,000)+.0769(-10,000) = 114,615

Therefore, optimal action: retain

c. $P(E|N) = \frac{(.2)(.6)}{(.2)(.6)+(.9)(.4)} = \frac{.12}{.48} = .25$, P(not E|N) = .75

d. EMV(S) = .25(50,000)+.75(50,000) = 50,000

EMV(R) = .25(125,000)+.75(-10,000) = 23,750

Therefore, optimal action: sell

e. yes

18.30 a. P(2 | 10%) = .01, P(1 | 10%) = .18, P(0 | 10%) = .81

b. P(2 | 30%) = .09, P(1 | 30%) = .42, P(0 | 30%) = .49

c. Probability of the states of 10% defective and 30% defective given:

# defective	10% defect	30% defect
2 defective	.308	.692
1 defective	.632	.368
0 defective	.869	.131

EMV of actions	check	Do not check
2 defective	17,616*	14,464
1 defective	18,264*	17,056
0 defective	18,737	18,952*

*optimal action given the circumstance

18.32 a. Perfect information is defined as the case where the decision maker is able to gain information to tell with certainty which state will occur

b. The optimal action: Low risk stock fund (see question 18.12)

EVPI = .2(6600 – 4300) + .5(0) + .3(1200 – (-600)) = 1000

18.34 Given that the optimal action is: New center

EVPI = .4(0) + .4(70,000 – 60,000) + .2(90,000 – (-10,000)) = 24,000

18.36 The expected value of sample information is $\sum_{i=1}^{M} P(A_i)V_i$ where $P(A_i) = \sum_{j=1}^{H} P(A_i \mid s_j)$. For perfect information, $P(A_i \mid s_j) = 0 \; for\ i \neq j\ and\ P(A_i \mid s_j) = 1 \; for\ i = j, thus\ P(A_i) = P(s_i)$

18.38 EVSI = .3(345,000 – 280,000) + .31(395115 – 383815) + .39(0) = 23003

18.40 Given that the optimal action: retain the patent (see question 18.28)

EVSI = .42(0) + .52(50,000 – 23,750) = 13,650

18.42 a. EVSI = .11(-600 – (-910)) + .89(0) = 34.1
b. EVSI = .013(-600 – (-1540)) + .194(-600 – (-825)) + .793(0) = 55.87
c. The difference = 18.77
d. None
e. 24.75

18.44 a.

Payoff	-10000	30000	60000	70000	90000	13000
Utility	0	35	60	70	85	100

b. EU(New) = .4(100) + .4(60) + .2(0) = 64
EU(Old) = .4(35) + .4(70) + .2(85) = 59
Therefore, the optimal action: New center

18.46 94000p – 16000(1-p) = 0 ➔ p = 16 / 110

Payoff	-160000	0	94000
Utility	0	160 / 110	100

$$\text{Slope}(-16000,0) = \frac{160/110}{16000} = .00009$$

$$\text{Slope}(0,94000) = \frac{100-160/110}{94000} = .00105$$

Therefore, the contractor has a preference for risk

18.48 a. P(S1) = .3(.6) = .18, P(S2) = .42, P(S3) = .12, P(S4) = .28
b. EMV(A1) = 460, EMV(A2) = 330, EMV(A3) = 0, EMV(A4) = 510
Therefore, the optimal action: A4

c. Draw the decision tree

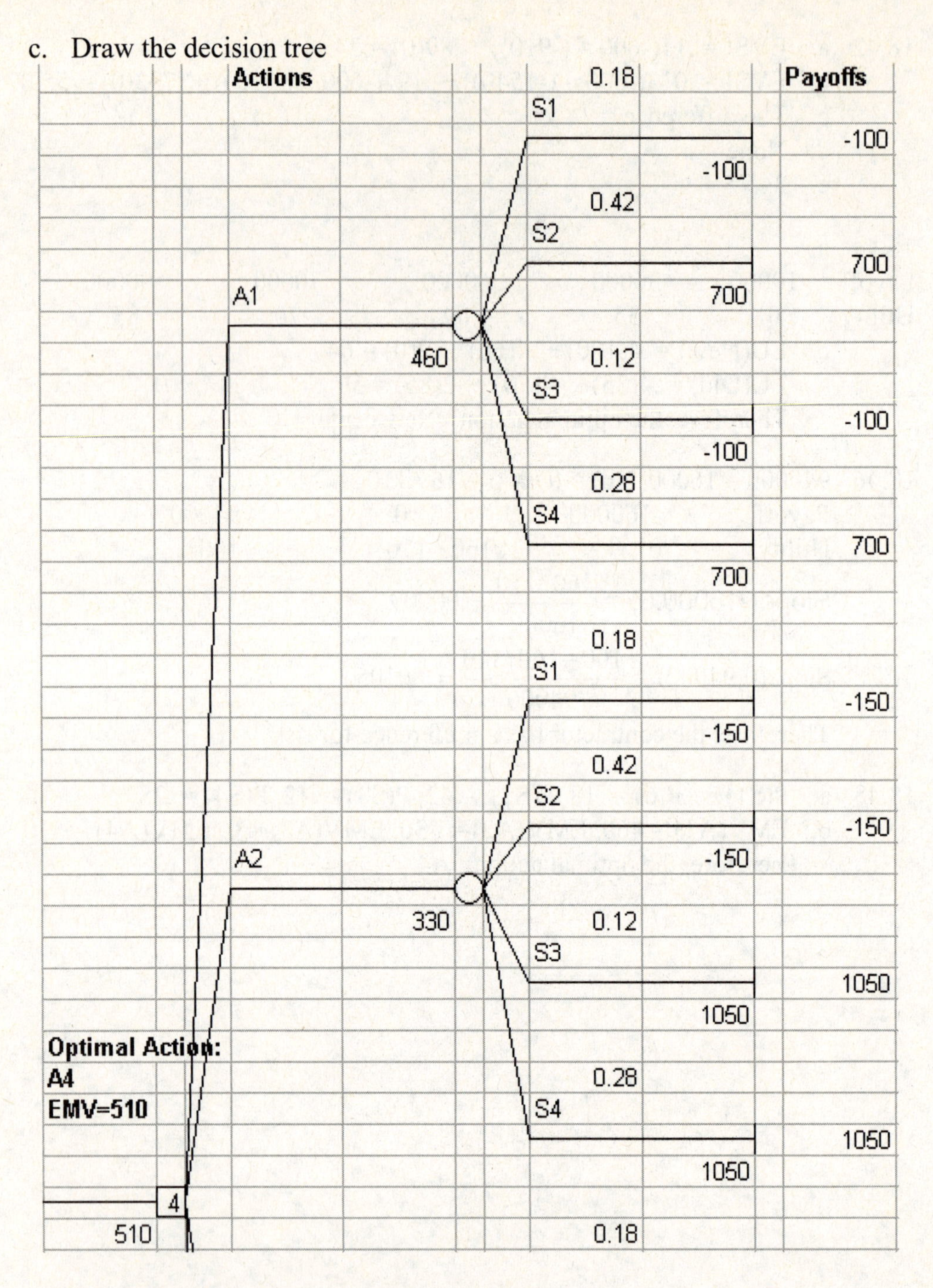

(Continued on next page)

TreePlan (Continued for 18.48):

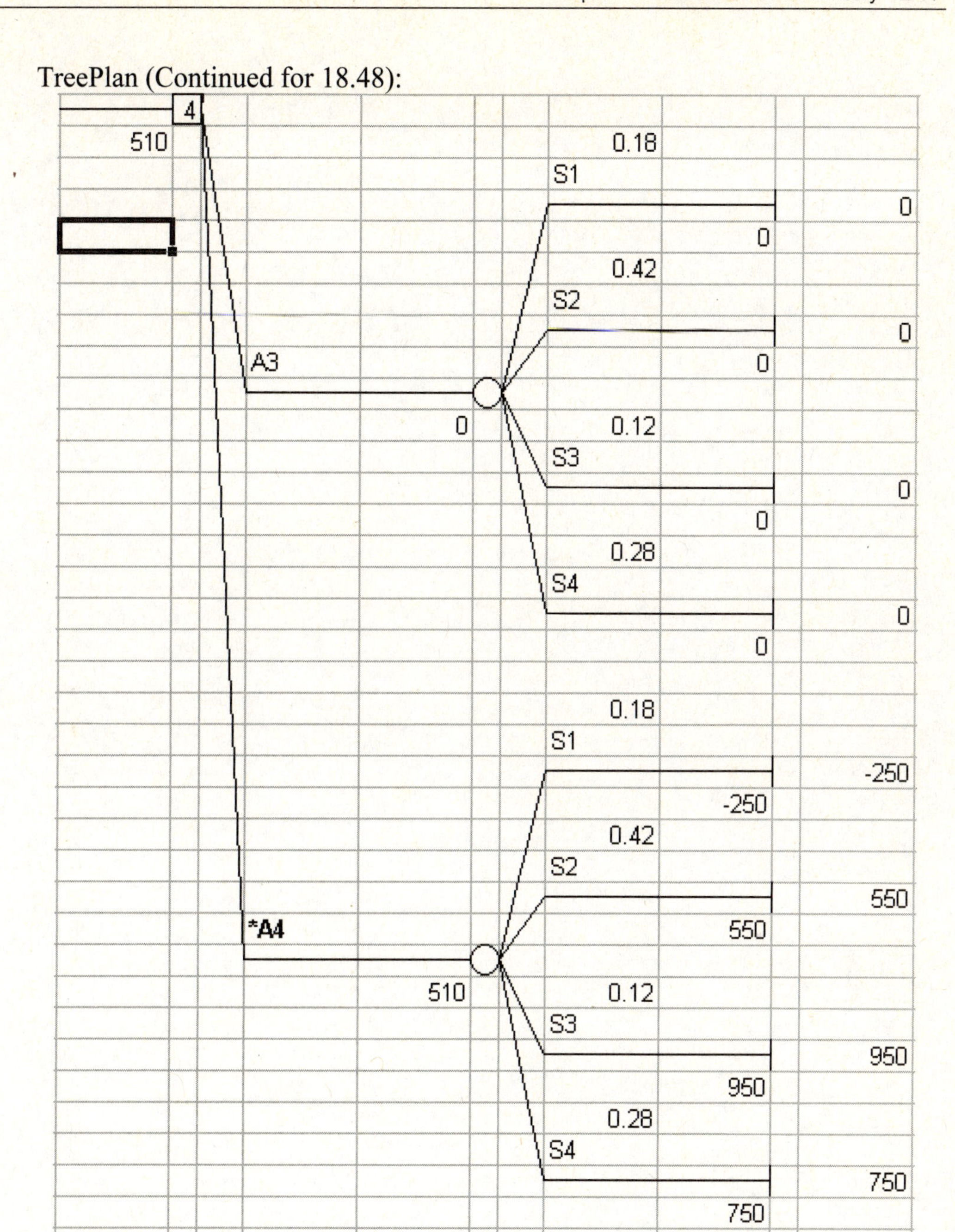

d. EVPI = .18(250) + .42(150) + .12(100) + .28(300) = 204

e. 79